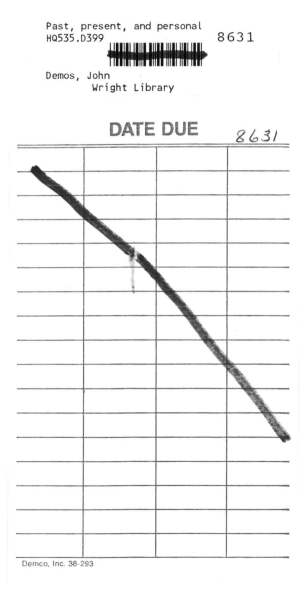

DATE DUE   8631

PAST, PRESENT, AND PERSONAL

Past, Present, and Personal

The Family and the Life Course in American History

JOHN DEMOS

New York   Oxford
OXFORD UNIVERSITY PRESS
1986

Oxford University Press

Oxford   New York   Toronto
Delhi   Bombay   Calcutta   Madras   Karachi
Petaling Jaya   Singapore   Hong Kong   Tokyo
Nairobi   Dar es Salaam   Cape Town
Melbourne   Auckland

and associated companies in
Beirut   Berlin   Ibadan   Nicosia

Published by Oxford University Press, Inc.,
200 Madison Avenue, New York, New York 10016

Oxford is the registered trademark of Oxford University Press

Library of Congress Cataloging-in-Publication Data
Demos, John
  Past, present, and personal.
  Includes index.
  1. Family—United States—History—Addresses, essays, lectures.
2. Family—United States—Historiography—Addresses, essays, lectures.
3. Public history—United States—Addresses, essays, lectures.   I. Title.
HQ535.D42   1986   306.8′5′0973   85-30996
ISBN 0-19-503777-4

1   3   5   7   9   8   6   4   2
Printed in the United States of America
on acid-free paper

In Memory of Philippe Ariès,
Father of Family History

Contents

Introduction

This is a book of gatherings from an historian's garden. The fruits are of somewhat variable size and shape, but they have all been grown in the same plot, if not precisely on the same tree. Childhood history, adolescence history, the history of middle and old age; the history of attitudes within (and towards) domestic life, of fatherhood, of social policy toward children: add an overview of the larger "field" (placed first in what follows), and you have the full list. Each has ripened within the past seven or eight years. All come under the general rubric of what is now called "family history."

The seeds were planted, innocently enough, long ago. I was a student, newly arrived in graduate school, groping to find a topic for a research seminar. My professor put the obvious question: "Well, what sort of thing are you interested in?" There was a pause; then, for reasons I cannot now recall, I mumbled the word "family" and asked whether that might make a "legitimate subject of historical study." The professor's response was to open his desk-drawer, remove a small folder of typescript pages, and toss it in my direction. "Here," he said; "this paper came just the other day from some people in England. It's a bit of family history. Why don't you look it over; maybe you could try something similar." Family history: how odd the term sounded then. But, correctly applied—the paper was indeed about family life in the past, one of the first such attempted anywhere. And I knew at once it was for me.

There is no need now to describe all the intervening parts of a long and rather peripatetic career; and this particular harvest is, in fact, considerably removed from those beginnings. Suffice to say that my earliest labors in family history were strictly by way of academic research. But later I began, without quite realizing it, to try something else—"public history," as some have called it—discussion about history, aimed at a wider audience. Wider than what? Than my fellow-historians and students. Meaning whom? Scholars from other disciplines, practitioners in other professional fields, and also—dare I say it?—that ultimate target of authors and booksellers, "the educated general reader."

The eight essays gathered here are all "public history." Each was conceived as a project of outreach to non-historians and/or non-scholars. (Seven were associated with specific invitations received from outside the bounds of my discipline.) Each was presented first as a lecture; the auditors have run a wide gamut from sociologists and anthropologists, to psychologists and psychiatrists, to "humanists" of several sorts; from lawyers and businessmen, to social workers, policy analysts, and assorted "citizen volunteers." Four have been previously published, always in collections of an explicitly inter-disciplinary nature.

These audience considerations have influenced substance, style, and tone of the whole batch. Family history has become, in recent years, a highly sophisticated—in some parts technical and downright arcane—area of research. Going through the latest journal articles, for example, can be a harrowing experience. However, the essays in this volume are decidedly non-technical—and, I do hope, non-harrowing to read. I have tried throughout to give ideas and argument top billing; I use evidence more to illustrate, than to prove, a given point. I have also sought to bar the door to technical language (and to explain it when it nonetheless sneaks through). At very least these writings should be *accessible* to all who care to look at them.

Some advance notice, too, about the limits of their "coverage":

People: "*The* family"? And "*the* life course"? Literally construed, such terms are a fiction. No two families, no two lives, can possibly be the same—let alone a hundred, or a thousand, or a million. The task, therefore, is one of averaging, of finding what statisticians would call a "central tendency." But this, in turn, is frankly discriminatory: cases that lie far from the average will be poorly represented, or not represented at all. There is no denying the final result: I have written mainly about people who were white, who were not identifiably "ethnic," who were "middle class." I regret not being able to do more, but see no need to apologize. The limits of the essays are the limits of the sources, and perhaps of my own imagination as well. Besides, few students of American culture would deny its centripetal aspect—the way those on the margins move, over time, toward the middle. The concept of "central tendency" is thus doubly useful here: it has dynamic as well as descriptive connotations.

Place: The essays deal, by and large, with things American— with families, and lives, and events, that belong to the history of the United States. Scholarly research on the family has been quite impartial—some might say imperious—in scattering its bless- ings world-wide; but my interest, and competence, is that of an "Americanist." Readers with a preference for European history should thus be forewarned.

Time: I have dealt casually, perhaps crudely, with the dimen- sion of time. Readers will have to search these pages high and low to find a specific date or year. Instead, I resort to decades, or whole centuries, or even to such expansive categories as "modern" and "pre-modern." (The latter terms demand at least a one-line defini- tion: I use them in reference to an imaginary boundary set at roughly the year 1800.) However, this is not to say that time is of little account here—quite the contrary. With one exception, these essays are designed to *feature* change over time, to compare and contrast developments between *different* chronological periods.

Such comparison has, indeed, the greatest importance for me; it goes to the core of historical understanding. And it touches a last, key point that merits full exposure in this introduction. I have called the book *Past, Present, and Personal*. The "past" part is evident enough; most of the content comes from time prior to our own. The "present" part may seem, at first glance, more question- able. But consider: Can we know the past, can we even approach the past, by any route save that which leads *back* from our own experience? Philippe Ariès said it bluntly and best on the first page of his magnificent, subject-opening study *Centuries of Childhood*:

> I must make it clear from the start that there is no question here of a gratuitous examination of society under the ancient regime. I have explained elsewhere how difficult it was for me clearly to distinguish the characteristics of our living present, except by means of the differences which separate them from the related but never identical aspects of the past. Similarly I

can tell the particular nature of a period in the past from the
degree to which it fails to resemble our present. This dialectic
of past and present can be fairly safely neglected by historians
of "short periods," but it must be used in the study of manners
and feelings whose variations extend over a "long period."
This is the case with the family, with day-long, life-long
relations between parents and between parents and children.

Other historians surely feel "this dialectic of past and present," but
few acknowledge it openly. Some, indeed, seem bent on keeping it
a secret—as if it might compromise the truth of their endeavors.
Here is the scholar's old fetish of "objectivity," reduced at its outer
edge to self-obliteration.

It is a myth, and there is no shame in saying so. In fact, history
without a "dialectic" would be pointless (except, perhaps, as sheer
escapism). We study the past for many reasons—not least because
we wish to know *better* our present. And this knowing lies, above
all, in the element of contrast: what was *versus* what is, the differences, the similarities, and all the shadings in between.

The passage from Ariès further suggests that these contrasts are
especially important for certain kinds of history: those that concern "manners and feelings," "the family" and its "day-long, life-
long relations." Which leads to the third term in my title—the
"personal." A scholar whose focus is political institutions, or
diplomacy, or official life more generally, may—just possibly—
succeed in avoiding the above-mentioned dialectic; at least he may
seem to avoid it. (I have my suspicions even there.) But for a
historian of personal life there is no such possibility at all. There is
no alternative to starting out from what he knows at first hand.

And where the past is personal, so, too, is the present. Historians can no more represent the whole of their own time than they
can detach themselves from it. The present is, ultimately, the
present I see—and feel—and understand. To be sure, my aim is
always to connect with others, to try to know their present, too,
"from the inside." The dialectic does not impose a full-blown
subjectivism; but, to some extent, we historians are inevitably
studying ourselves.

Again, it seems best to say so out loud. Here is another nugget
from Ariès, this one from the start of a chapter on "Children's
Dress."

As soon as the child [of the Middle Ages] abandoned his
swaddling band, . . . he was dressed just like the other men
and women of his class. We find it difficult to imagine this

confusion, we who for so many years wore knickerbockers, the now shameful insignia of retarded infancy. In my generation we came out of knickerbockers at the end of the fifth year at school; my parents, urging me to be patient, quoted the case of an uncle of mine who was a general and who had gone up to the military academy in knickerbockers.

Another contrast—in which the historian brings himself, his own experience of "the present," right to center-stage. And how striking, how resonant it is! There were no knickerbockers in my own childhood, but I do recall my feeling about "short pants"—and the great leap forward to trousers.

A reader will shortly see that I have followed the same course. If my "present" and "personal" are not identical to his (hers), neither can they be entirely different. Here is a second dialectic which historical work should also encourage, one that brings writer and reader into a kind of *exchange*. In fact, from the reader's standpoint, there is a multiple exchange here: my present and yours, my past and yours. This should be an opportunity on both sides.

PAST, PRESENT, AND PERSONAL

CHAPTER I

Digging Up Family History: Myths, Realities, and Works-in-Progress

The opening essay in this volume was—and is—designed to introduce the field of family history. First prepared as an invited address for the 1982 annual meeting of the American Psychiatric Association, it assumes no prior knowledge of the field, and surveys a huge expanse at a high rate of speed. It proposes, in fact, a kind of "map" of the territory—admittedly one of many that might be found, and used, by a traveler bent on entry there. Scholarship being what it is, no two map-makers could possibly arrive at the same result.

My map purports, first of all, to distinguish the main sectors of the field, and then to trace the landmark features within each one. The approach is frankly "historiographic." Discussion proceeds by reference to work already performed—books (and articles) written, questions raised, documents discovered, controversies begun, conclusions reached—with some attention, as well, to the proverbial Need-for-Further-Research. Readers without a taste for such discussion may wish to skip straight to the substantive chapters that follow; still, in some respects this first piece anticipates all the others.

Indeed, I have not suppressed my own sense of the substance even here. And I do mean "my own sense." I am acutely conscious of holding a minority view about several key parts of this subject (for example, the emotional side of pre-modern family life), and I do not wish to belittle the alternatives. I have tried at least to identify the latter (if not in the text, then by way of a footnote), and readers may well decide to examine them later at their specific points of origin.

"Historiographic" or not, this essay was, in the doing, a particularly vivid experience for me. I had never before tried to write about the enterprise of family history, only to write in it. The APA invitation felt, then, like a call to describe a place where I have lived for many years. It is not always easy, in such cases, to keep an appropriate distance. And a reader may sense, at numerous points, my abiding affection, my occasional irritation, my deep-down sense of belonging just here—in short, the full range of feeling one inevitably has for the place one calls home.

This essay is a "report from the trenches" in a field of research that was unexplored, not to say unimagined, as little as two generations ago. The field is (what has come to be called) "family history." The trenches are as yet only half-dug; hence they are shallow in many spots, ragged along most of their edges, and littered with loose dirt throughout. But rough and unfinished as every part of this seems, it is possible to discern the contours, the proportions, the design of the whole. And since the work is very much ongoing, it is reasonable to look for further progress just ahead. In fact, the field *and* the trenches are swarming with eager diggers, whose enthusiasm alone must count for something. The diggers, of course, are professional historians, drawn to the lure of new discoveries. They dig for themselves and for each other—but would be pleased to interest a wider audience.

I

For a territory so recently brought under investigation, no one "map" can be regarded as definitive. But the map to be presented here aims, nonetheless, for general coverage. And it does so by way of four principal sectors or subdivisions—each of which merits separate consideration.

The first sector one comes to—at least the first that family historians came to—is clearly marked as *demography*. Work in this area began some three decades ago and has continued without pause to the present day. The initial advance came from a pioneering band of French demographers associated with the so-called *Annales* school.[1] Their lead was soon communicated to a group of

English scholars—the Cambridge Group for the Study of Population and Social Structure—and then, as the news spread further west, across the Atlantic to the Americas.[2] As a result we now know a great deal about the demographic contours of family life in virtually every part of modern history, and not a little about the *pre*-modern situation as well. We know, for example, all about "mean household size"—MHS in the argot of the Cambridge Group—through a full four centuries of English history. The gist, in that connection, is a remarkably stable mean of 5-6 persons per household until the onset of the Industrial Revolution, and a very gradual decline thereafter.[3]

Actually, mean household size is not by itself a very interesting *datum*, except, perhaps, as it serves to confute a stubborn piece of historical mythology. It is still rather widely believed that "extended families" were the predominant form until just a century or so ago. But if by "extended" we mean three (or more) generations, including married siblings and their various children, all living under one roof—that has *not* been the shape of Western family life for as far back as the records allow us to see.[4] Instead, the simple "nuclear" unit (husband, wife, and their natural-born children) has always been with us—admitting only of limited add-ons (a servant, an apprentice, an aging grandparent or two) in particular times and places.[5] To be sure, many pre-modern communities supported a density of kin-contacts unknown in our own day; for siblings and cousins and in-laws frequently did inhabit the same village or parish neighborhoods.[6] But not the same *households*. Thus if co-residence is the defining factor, the concept of "extension" seems misapplied.

Mean household size is the net product of other demographic indices: age of marriage, frequency of *re*marriage, rates of fertility and mortality. And the latter hold much interest in their own right. They also hold some surprises for conventional wisdom in this area. The present writer can recall his own surprise, in a graduate seminar-project twenty years ago, upon discovering that most New England colonists of the seventeenth century married in their middle-to-late twenties. (That range has, in fact, been found typical for many other pre-modern populations as well.) He can also recall his astonishment at the sheer longevity of his "Puritan" study-sample: nearly seventy years, on the average, for those who survived the special health hazards of early childhood.[7] (In this respect New England turns out to have been very remarkable; most contemporaneous settings supported a far lower standard.[8]) Other results have been *un*surprising, though not unimportant: for example, high rates of fertility (eight to ten completed pregnancies per married couple), and thus (where mortality was not equally high) large complements of children.[9]

These summary comments cannot possibly do justice to the richness and complexity of recent demographic research. For, in addition to establishing the quantitative boundaries of family life, demographic analysis has thrown light into some of its most private corners. For example, we can now discover, from evidence of birth-spacing and "age-specific maternal fertility," the point at which given populations first began to practice birth control.[10] This development—a veritable sea-change in family history—is inaccessible to all other methods of study, since the people involved would not ordinarily document their practice in such intimate matters.

Much of the demographic work raises questions of comparative—that is, trans-historical—psychology. To wit: (1) What are the implications, for psychological life, of pre-contraceptive attitudes? (Perhaps a lessening of certain internal pressures and conflicts?) And (2) what is the impact, on personality development, of growing up in a very large family? (More, or less, or simply different "sibling rivalry"? A diluted experience of the "oedipus complex"?) And (3) how might "separation" issues be construed in settings where death is a common experience for people of all ages and every social station? (A distinctive range of "mourning reactions"? A certain degree of "psychic numbing"?)

These last are questions only, for which demography cannot by itself supply direct answers. There has been, in fact, a tendency among enthusiastic practitioners of the demographic art to leap from numerical *quantities* to psychological *qualities*. Thus, for example, high levels of early-life mortality have led some scholars to infer low levels of parental concern for infants and young children. The implicit—sometimes explicit—premise is that parents would not allow themselves to become much attached to offspring whose propects of survival were at all in doubt.[11] Similarly, rates of illegitimacy and of what demographers call "bridal pregnancy" have been taken as a measure of power relations within the family. In certain communities of eighteenth-century America, 30 to 40 percent of all brides were going to the altar pregnant: that much one can demonstrate by comparing their wedding-dates with the birthdays of their eldest children.[12] But can one then assume—as some scholars *have* assumed—an absence of "control" by the older generation, and a high degree of "auton-omy" in the younger one?[13] The factors involved here are too numerous, and the relationships between them are too complex, to permit such straightforward efforts of inference. And they suggest a larger point as well: demographic study is best seen as setting the stage—not writing the script—for the vital inner dramas of family history.

II

It is to the inner dramas that we turn next. The division of power, the demarcation of roles and responsibilities: thus the family in its "structural" aspects. Here, too, history unfolds a long and change-able story, which historians are only beginning to understand. The concept of "patriarchy" has served as an entering wedge. In pre-modern times—so the argument frequently goes—*fathers* ruled families with a more or less iron hand. Later (roughly the nine-teenth century) their grip was progressively loosened by all the trends of economic and cultural modernization.[14] Most scholars would be willing to accept this very rough model of historical change, but the particulars are so enormously variable that the model by itself does not mean a great deal. "Patriarchy" in relation to whom? and by what means? and to what specific ends? And how does one measure "power" and "responsibility" in the first place? Such questions suggest a need to disassemble the "family struc-ture," the better to see its constituent parts. One can study the marital pair—husband and wife—as a structure of power in its own right. One can study parents and children as another kind of structure, with further refinements that distinguish fathers and mothers, daughters and sons. One can even bring grandparents and grandchildren into view, where there are long-term issues of "lineal descent."

But with the structural lines at least roughly drawn, it remains to identify the substantive contexts in which power relations are most fully expressed. Here, too, there is much room for historical flux and change. For example: in many pre-modern communities inheritance was a vital nexus of power in the family. Fathers might try to constrain the behavior of their maturing children by grant-ing (or withholding) family properties.[15] Unfortunately, it is not enough in these matters to study the set phrases of wills and other probate documents; a man who is left out of his father's will may be unusually rich or rebellious, or may simply be living far away.

In order to "read" such materials aright, other evidence is needed—evidence on the particulars of that family's experience.

A second fulcrum of power between the generations is—or, at least, was—the issue of "mate selection." Typically, in pre-modern times, parents were much involved in the courtship and marital decisions of their grown children. But the picture is easily over-drawn. Few children anywhere were obliged to marry against their personal inclinations; indeed, positive inclination was seen as one of the requisites of a good marriage (but not the only one). In some pre-modern families parents would initiate a match and seek to bring it to fruition, while the young people involved might exercise a kind of veto; in others, the young initiated, and the parents vetoed.[16] As time passed, the balance moved strongly in favor of the second pattern, and eventually parental influence might be circumvented altogether.[17] Here, surely, was a "power-shift" of great historical consequence.

And yet any such change must be described with caution. The closer one gets to the details of power relations within the family, the more complicated—and the less amenable to summary formulas—they come to seem. This sounds obvious enough, when stated in such general terms, but it merits underscoring in a specific historical case. Consider, then, John Dane—born in England in the opening years of the seventeenth century, son of a tailor and himself trained as a tailor, an altogether ordinary specimen of pre-modern humankind. The only *extra*ordinary thing about him is a little autobiographical memoir, written near the end of his life and fortuitously preserved to our own time. Thus we have, in his own words, some revealing information on his relation to his parents.[18]

Being . . . about eight years old, I was given much to play and to run out without my father's consent and against his command. Once when my father saw me come home, he took me and basted [i. e. beat] me . . . My father and mother . . . told me that God would bless me if I obeyed my parents, and what the contrary would issue in. I then thought in my heart—oh, that my father would beat me more when I did amiss!

When I was grown to eighteen years of age or thereabouts, I went to a dancing school to learn to dance. My father, hearing of it . . . told me that if I went again he would baste me. I told him that if he did, he should never baste me again. With that my father took a stick and basted me. I took it patiently and said nothing for a day or two, but one morning betimes I rose and took two shirts on my back and the best suit I had, and put a Bible in my pocket, and set the doors open, and went to my father's chamber door, and said, "good-bye, father; good-bye, mother." "Why, whither are you going?" "To seek my fortune," I answered. Then said my mother, "go where you will, God will find you out." This word, the point of it, stuck in my breast; and afterward God struck it home to its head.

I thought my father was too strict—I thought Solomon said "be not holy overmuch," and David was a man after God's own heart and he was a dancer. . . . [And so] I went on my journey and was away from him half a year before he heard where I was. (*There followed a period of some two years when John Dane practiced his trade, on his own, in several different English towns.*) . . . But at last I had some thoughts to go home to my father's house; but I thought he would not entertain me. But I went, and when I came home my father and mother entertained me very lovingly.

(*In time he left again, and married—no mention of consulting his parents about this—and settled with his bride in still another part of the country. And then, in about 1635, he made the fateful decision to migrate to New England.*) When I was much bent to come, I went to my father to tell him. . . . My father and mother showed themselves unwilling. I sat close by a table where there lay a Bible. I hastily took up the Bible, and told my father that if where I opened the Bible, there I met with anything either to encourage or discourage, that should settle me. I opening of it . . . the first line I cast my eyes on was: "Come out from among them, touch no unclean thing, and I will be your God and you shall be people." My father and mother never more opposed me, but furthered me in the thing, and hastened after me as soon as they could.

How shall we evaluate this little *vignette* as an instance of the politics of family life long ago? On the side of "patriarchy" we may count (1) the father's readiness to "baste" his young son for misbehavior (and the son's acceptance of that procedure); (2) the father's assumption that he can prevent his son (even as a youth of 18 years old) from learning to dance; and (3) the son's assumption (though by this time he was married and a fully independent tradesman) that he must obtain his parents' consent before moving to America. On the other side—against patriarchy—there is (1) the son's pursuit of his own aims and wishes from a very early age; (2) the son's declaration of independence when (again, at age 18) he leaves home to "seek his fortune"; (3) the clever way in which the son maneuvers around his parents' opposition to his plan of removal to New England. (To be sure, he does get timely assist from the Bible!) Where should the greatest emphasis be placed? This is a question on which reasonable scholars may well disagree.

In fact, power is an especially problematic issue, given the always imperfect record of centuries past. Role, by contrast, is much easier to study. Usually there are some prescriptive materials (such as sermons and advice books) to indicate the predominant values, and sufficient personal documents to exemplify behavior. Consider productive life, for example. Throughout the pre-modern world family-members (save only the youngest) produced for their common good, in visibly direct and meaningful ways. Of course, most pre-modern families were rooted to farms—a situation which even today promotes a good deal of work-sharing.[19] In early America there was some division of labor: men in the "great fields," plowing and planting; women in the orchards and dairies, or indoors by the hearth; older children helping out as needed, mostly with their same-sex parent. But each could appreciate— could *see*—the contributions of the others; and all could feel the underlying framework of reciprocity. Moreover, in certain "crisis" periods, like the harvest, all worked side-by-side for days on end. With the advent of urbanization and modern commerce and industry, this framework broke apart. Men were now "providers," "breadwinners," producers *par excellence*.[20] Women were literally domesticated; their role became that of "homemaker" in a newly exclusive sense.[21]

These changes carried, in turn, important implications for child-rearing. In the colonial period the primary parent had been *father*. Books of child-rearing advice had been addressed to him; the law had preferred him (to mother) in the matter of child custody; and all parties affirmed his superior "wisdom" in understanding and nurturing the young. (Woman were considered too irrational and unsteady to take the lead here.) In the nineteenth century the pattern was rapidly, and radically, altered. Father's obligation to "provide" left him little time and energy to nurture (at least in personal ways). Mother's duties, meanwhile, became virtually all-encompassing. ("All that I am I owe to my angel-mother": thus a favorite period cliché.) The same sort of role-separation has, of course, continued largely intact to the present day. But with some new winds beginning just now to waft through American family life, it may be instructive to remember the earlier period.[22]

The topic of family structure raises, inevitably, fundamental questions of gender. And gender itself makes a lively center of study. Indeed "women's history" (so called) has become one of the most lively of all research sub-fields in the past ten or fifteen years. The substance can be sampled in a plethora of books and articles; the spirit is evident in some of their titles: *A Heritage of Her Own*,[23] *Clio's Consciousness Raised*,[24] *The Majority Finds Its Past*,[25] and so on. But although the spirit is broadly feminist, it is not, for the most part, overtly polemical. And the resultant gains for historical understanding are truly of the first importance.

The earliest forays in women's history—antedating the recent upsurge—expressed what now seems a rather narrow concern with politics. The origins of organized feminism; the struggle for the suffrage; the activity of women reformers more generally: such were the leading questions.[26] But as one work led on to another, the boundaries widened—and the questions deepened. There emerged a rough outline of women's past experience—at least, American women's past experience—which went approximately as follows. In colonial America women were defined as being inferior to men ("He for God, and she for God in him"), but, practically speaking, they retained considerable scope for initiative and self-assertion. As previously noted, they made important contributions as farmers' wives and daughters; and some of them also worked as tradespeople, physicians, innkeepers, and the like. They could not vote or be otherwise active in public affairs; but their influence was exerted, was *felt*, in countless informal ways.[27] Then came the nineteenth century, with its massively transformed human ecology—and, for women, a severe loss of status and function.[28] "Homemaking" proved to be a form of domestic imprisonment; polite culture was sheer vapidity; public life remained (more than ever) off limits. And when, at mid-century, organized feminism was born, it expressed an anguished cry from the depths of oppression. The plot-line of women's history ever since has been a stop-and-go effort to escape from those depths—or, stated in less extreme terms, to push back the limits of constraint.

But this historiographic overview is now being reconsidered, and at least partially revised. Its advocates are charged with having fabricated a "golden age" of women's history (the colonial period), and, concurrently, with having ignored certain positive features of the succeeding period.[29] Indeed, recent scholarship very nearly reverses the balance. Colonial women are still seen as active, to be sure, within the family and elsewhere—but always under the heavy dominance of men. Nineteenth-century women had, at least, a *place* that was primarily theirs (home), and a *vocation* that was theirs alone (child-rearing); hence, within some confines, they knew "autonomy"—and made the most of it.

The debate, thus briefly summarized, is as yet far short of resolution. However, it seems necessary to stress that "autonomy" has both a social and a psychological aspect. Historians, unfortunately, are not always clear about the distinction; too often we stress the former, while ignoring (or misunderstanding) the latter. Indeed, this whole territory quivers with subterranean resonance, some of which links women's history directly with psychiatric history. The first important cohort of psychiatric patients—including Freud's early patients—was composed of troubled women from the "comfortable classes" of the late Victorian era. "Hysteria," "neurasthenia," "breakdown": whatever the favored diagnostic category, their symptoms reflected strikingly on their life-situation. The prevalent cultural values declared that women should be submissive, selfless, ceaselessly effective on behalf of others ("ministering angels," in the idiom of the time). But these women, in their illness, managed to be *in*effective, self-absorbed, and tacitly dominant over friends and family members. The illness, in short, seemed to mock the cultural values, and may plausibly be viewed as a form of protest.[30]

Many of these patterns and tendencies can be organized around the concept of "identity." Simply put, women's identity and men's seemed to diverge so radically in the nineteenth century that all human communication across the gender-boundary was impaired. Experience differed, of course (again the work/family dichotomy); but feelings, intelligence, "sensibility," and "moral inclination" also differed in elemental ways. As a result, courtship was heavily burdened with anxiety and doubt, and women, in particular, underwent a characteristic "marriage trauma."[31] There must have been redeeming, even rewarding, aspects of Victorian marriage in many individual cases, but they are not easily discovered from the vantage point of a century later. Sex was one part of the problem—cultural standards defined women as "passionless," and the standards were sometimes enforced by the surgical procedure of clitoridectomy—but it was not the only part.[32] Women expected "understanding" and "sympathy" only from other women; together they created what one historian has called a "female world of love and ritual."[33] Men did something similar in their own world (substituting "animal energy" for love); and social experience of all kinds seemed to divide on same-sex principles.[34]

These male/female questions lead on endlessly, and there is no need in the present context to follow their trail any further. Instead, we must turn to age and aging—a topic of equally large importance. For every family is (and was) both a system of gender relations *and* a system of age relations. Power, status, and responsibility within the family are defined by the second no less than the first.

But in order to put age relations in historical perspective one has to know how the people involved have defined and demarcated the aging process. The verdict of scholars in this area, while still very tentative, has at least an underlying coherence. In general, history has brought a greater and greater differentiation of the life course, and a sharper experience of its constituent parts. One of the earliest and most influential books in the entire literature of family history presented a remarkable picture of childhood over time.[35] In pre-modern society childhood was barely recognized as such; young people appeared chiefly as "miniature adults." Only in the seventeenth and eighteenth centuries—and then only in elite groups—did children begin to receive special consideration for their distinctive needs, interests, and vulnerabilities. By some accounts adolescence is an even more recent invention. In fact, it was a famous American psychologist, Professor Granville Stanley Hall, who put adolescence on the map (so to speak) of modern popular culture. But these developmental categories themselves expressed fundamental changes of experience. And experience is what counted most of all. To some extent, modern adolescence expressed an altered balance of social circumstance—the decline of apprenticeship, the growth of mass public education, the development of new living-situations for young people exiting from their families of origin. But there was also an innerlife aspect—the growth of "identity diffusion" (in the psychoanalyst Erik Erikson's terms) in the face of ever-widening life-*choices*.[36]

Historical studies of childhood have necessarily underscored the matter of child-*rearing*. We have already noticed, in another connection, the shifting balance of responsibility as between fathers and mothers. But what of the goals and methods of child-rearing; and, finally, what of its outcomes? Historians have been much intrigued by these questions, too—though somewhat frustrated by the limits of the written record. Often enough one finds abundant evidence of *prescription*, but precious little of *practice*.[37] Nonetheless, certain broad trends do come clear. In early America child-rearing was framed by the doctrinal imperatives of evangelical religion. "Original sin" was associated with all of humankind, not least with its youngest specimens. Infants came into the world as carriers of a "diabolical" tendency, and right-thinking parents would react accordingly. The tendency was identified with "pride" and "self," and above all with "will." Thus the advice-literature urged the "breaking" and "beating down" of will—those were the favorite verbs—from the earliest possible age.[38]

It took many generations for such advice to disappear altogether; but by the mid-nineteenth century the emphasis—and, most especially, the tone—had shifted. Children of that era were viewed as being morally neutral, even "innocent." "Nurture," not will-breaking, became the touchstone of the advice-literature; parents were urged to mold their young by the "gentle arts of persuasion"—and by their own good example.[39] In fact, the ends of child-rearing had changed along with the means. There was a new and growing emphasis of qualities of independence, resourcefulness, initiative—a whole expressive mode. Only thus would a young person be prepared to seize the main chance and "go ahead" in the open society of modern America. At the same time, most authorities stressed the need for an inner "compass" that would hold behavior within morally acceptable bounds.

Translated into the language and concepts of our own time, the nineteenth century seems to have replaced an old child-rearing regime based on shame with a new one based on guilt.[40] Parents in the colonial period had frequent recourse to scolding, humiliation, and edicts of temporary banishment. The famous Puritan preacher Cotton Mather wrote as follows about his own practice: "The first chastisement which I inflict for an ordinary fault is to let the child see and hear me in an astonishment, and hardly able to believe that the child could do so base a thing . . . [And] to be chased for a while out of my presence I . . . make to be looked upon as the sorest punishment in the family."[41] Nineteenth-century parents, by contrast, stressed the hurt given to others—especially to themselves—by the child's misbehavior.

It is necessary to mention, if only in passing, scholarly interest in the later life-course. For example, recent years have brought a boomlet of studies on the history of old age.[42] Much of this itself reflects a historical phenomenon—the emergence, in the twentieth century, of old age as a "social problem." Gerontology and geriatrics, two relatively new "sciences" with policy, clinical, and research dimensions all their own, have perforce turned some heads toward the past. If old age is a social problem now, what was it in former times? Surely, much has changed. The numbers have changed, for one thing: the numbers of the elderly within the total population, and the numbers within any given "birth cohort" who can expect one day to *be* elderly. Social factors have changed as well. Retirement—a more or less abrupt detachment from the world of productive work—is now the critical marker of the aging process; there was no precise pre-modern equivalent. Pensions, nursing homes, social insurance: these, too, in most respects are recent inventions.[43]

The "problem" theme has served to tilt historical interpretation. If old age is difficult *now*, probably it was less so *then*. When the elderly were fewer, perhaps they were more valued—even "venerated."[44] When they were not retired, they may have felt themselves to be more useful and capable. These hypotheses quickly add up to an historical *schema* paralleling the "golden age" view of women's history. Fortunately, they are empirically testable, at least in their main parts; and the tests to date suggest much need for caution. Pre-modern old people probably *were* better off than their present-day counterparts, as to social and economic position; but they seem to have paid a price for this, in various forms of psychological disadvantage.[45] How one gauges the overall balance is more an intuitive than a scholarly judgement.

III

Moving right along, one arrives at a third major area of "trench-digging" within the larger field. It is not, to be sure, recognized as a separate area by some of the diggers themselves, and it does adjoin (or overlap) parts of the territory previously described. Yet it should be distinguished from the others as clearly as possible. And it should be identified, in bold letters, as "emotional experience"— or, more simply, as "affect."

Because of their failure to identify this area for what it is, historians have not studied it with much care—or effectiveness. Indeed, they have *assumed* and *inferred* more than they have studied. And what are their reigning assumptions? First, that affective experience was generally impoverished in the past; second, that families, in particular, knew only a fraction of the emotional rewards that *we* look for in our domestic life today; and, third, that individuals possessed little capacity for emotional sharing with others. Pre-modern marriage—so the argument goes—was a matter more of "convenience" and "instrumental advantage" than of loving care. And pre-modern child-rearing was full of indifference, callousness, and outright brutality. In this regard the reigning view is a "progressive" view: family life, over time, has (allegedly) been getting better and better.[46]

And yet there are strong grounds for skepticism about this—about the parts, and about the whole. Admittedly, it is not difficult to find evidence that appears—by the light of our own values—to support the progressive view. But do notice the qualifier: "by the light of our own values." A modern-day American may well be surprised by the formal style of pre-modern marriage ("I am, as ever, your affectionate Husband, B. Franklin").[47] And the same observer will probably be shocked by certain elements of pre-modern parenting (the calculated use of fear and shame, and the regular resort to corporal punishment).[48] But seen in the context of an entire cultural *Gestalt*—theirs, not ours—such behaviors made sense, and were not inconsistent with genuinely warm feeling. Perhaps the point becomes clearer if we reverse the roles of observer and observed. Some of our own practice might well look "unfeeling" to our pre-modern forebears: for instance, the way we consign infants to separate beds and cribs, hold them to "schedule feedings," and leave them sometimes to "cry through" a tantrum.[49] A further objection to the progressive view is its reliance on behavioral fragments culled virtually at random from a motley array of records. Often such fragments are themselves anomalous by the standards of their time (hence their appearance in the records). Again, the counter-example of our own time may be instructive: Imagine a study of modern American family life based on "cases" covered in big-city newspapers.

The study of emotional experience remains, then, an important challenge to historians—important, but difficult, and largely unsatisfying in its results to date. A new start may well be necessary, if there is to be sustained progress in the long run. Context must, in the first place, be much more cleanly handled than has typically been the case so far. And there are other points to consider as well. A critical problem in "emotion history" is the complexity—the sheer untidiness—of the data; time and again, evidence of affect comes to us wildly entangled with other things (ideas, opinions, behavior, not to mention elements of the setting in which affect is expressed). In order to counter this problem, special steps must be taken to hold emotion itself clearly in focus. In addition, scholars may need to adopt careful principles for distinguishing between *different* emotions; perhaps they may even decide to adopt one or another theory of emotion (borrowed, presumably, from academic psychology). And they will be further obliged to separate what one scholar calls "emotionology"—attitudes, values, and ideas about emotion—from actual emotional experience.[50] (Here is the old methodological bugaboo of prescription *versus* behavior, in a somewhat different guise.) None of this has happened yet; fortunately, however, there are signs of change just ahead.

IV

With demography, structure, and affect all noticed in their turn, one large territory yet remains. In a sense, all the issues discussed thus far have a point of reference internal to the family (its size and shape, its distribution of power and responsibility, its emotional qualities). And, in the same sense, what is left to discuss has an external reference: the family viewed from the standpoint of its individual members, on the one side, and of the larger community, on the other. How, in short, are we to construe the overall pattern of relations between the family and its constituency, as history moves along?

To ask this is to raise issues of "function," of "cost" and "benefit," and of "fit" between one unit of experience and all the others. The family not only *is,* the family also *does*: it has its own center of activity, at once unique and profoundly influenced from outside. Long before family historians appeared, family sociologists had developed a model of change-over-time, around precisely these questions.[51] The family, they suggested, had experienced a long-term "erosion" of function. Where once it had served a broad range of plainly "instrumental" needs, in modern society it was reduced to two elements: the "socialization" of children and the provision of emotional support to its adult members. Historians, for their part, have measured this model against the evidence and used it to generally good effect. The family of pre-modern times was indeed a hive of instrumental activity: of production (e.g. the "family farm"), of schooling, of worship, of medical practice, and of care for all sorts of "dependents" (orphans, elderly people, the insane, even criminals).[52] And the transition to modern times has, for certain, reduced this range dramatically. Public schools, hospitals, asylums, prisons now stand in *place* of the family at many points. To recognize this is not to belittle the importance of those family functions which yet remain. The heterogeneity of modern society at large, and the isolation of domestic space from other settings, has raised the stakes all around. Perhaps, then, child-rearing has become more vital and difficult than before, and emotional support within families has taken on an ever-greater urgency.[53]

Yet even as family historians have (by and large) endorsed and applied this interpretive model, they have sought to refine it in several ways. The model exaggerates, or at least oversimplifies, change. Recent studies have, for example, discovered much interaction between the nineteenth-century family and the work-place. Some early manufacturing establishments built up their labor force by way of entire family units—husband, wife, and children hired together. In other cases, the interaction was more circuitous. The famous "mill girls" of Lowell, Massachusetts, and elsewhere, were, in effect, the center of a broad-gauge human exchange—between rural families with labor to spare, on the one hand, and the factory-system, on the other.[54] Meanwhile, foreign labor was brought to America by a process known to scholars as "chain migration." Typically, a young man from Europe, the Far East, or Canada forged the first link in the chain by migrating and finding work; then he helped his relatives to join him in the same community—often as an employee of the same *company*. Thus, when studied from close up, the work force of particular factories seems a honeycomb of kin-based connections.[55]

As with industrial work, so, too, with other sectors of social experience: the family was a presence in all. Account books kept by merchants and tradesmen reveal a pattern of regular reliance on kin; likewise the rosters of urban political "machines," and even lists of converts in religious revivals.[56] The pattern should not be exaggerated, and the effects must not be construed as running entirely one way. The family was as much acted *upon* as it was an actor in its own right. Moreover, events in the long term surely worked to weaken most of these relationships, so that now only vestiges survive. Still, no adequate account of modern social history could leave them out altogether.

This, indeed, is a measure of the growing maturity of historical research on the family. In its first phase, such research was necessarily set apart from historical studies at large. The boundaries had to be surveyed, the benchmarks established, the basic internal topography understood, before any links to adjacent territories could be explored. But now the situation has changed. The links hold greater and greater interest, and not only for family historians but for scholars right across the board.

V

As noted at the outset, the enterprise of family history is very much ongoing. No single question can be considered as finally resolved, and new questions are popping into view all the time. Nonetheless the results so far have considerable shape and structure; and at several points they refute, or revise, well-known elements of conventional wisdom. Moreover, they set some limits around what all of us—whether historians, social scientists, or simply concerned citizens—may reasonably expect of our future. Is the family likely to transform itself into something radically new and different, or even to disappear entirely? No, *not* likely—given its impressive record of durability through many prior centuries. Might gender-roles so alter that parental and household responsibilities are no longer predominantly assumed by women? Quite possibly—given the variable patterns, in this matter, of the past. Will childhood itself wear a different aspect (e.g. more "hurried," as one recent prognosticator has put it) in times to come?[57] Yes, insofar as childhood responds to cultural circumstance; but probably no, where inner-life development is concerned.

To raise these questions is to spotlight the unceasing traffic between our past, our present, and our future. And it is also to conjure up another kind of traffic, one that crosses the borders of academic disciplines and finally confronts personal experience. From such imaginative journeys we may all hope to profit.

NOTES

1. For an early and important example of this work, see Louis Henry, *Anciennes Familles Genevoises* (Paris, 1956).

2. See Peter Laslett and Richard Wall, *Household and Family in Past Time* (Cambridge, Eng., 1972).

3. *Ibid.*, 125–203.

4. Peter Laslett, *The World We Have Lost* (New York, 1965), ch. 4.

5. For an early American instance of this pattern, see John Demos, *A Little Commonwealth: Family Life in Plymouth Colony* (New York, 1970). On the matter of "add-ons," and its implications, see Lutz Berkner, "The Stem Family

and the Developmental Cycle of the Peasant Household: An Eighteenth-Century Austrian Example," in *American Historical Review*, LXXVII (1972), 398–418; and Michel Verdon, "The Stem Family: Toward a General Theory," in *Journal of Interdisciplinary History*, X (1979), 87–105.

6. This point is explored at length, for one early American community, in Philip J. Greven, Jr., *Four Generations: Population, Land, and Family in Colonial Andover, Massachusetts* (Ithaca, N.Y., 1970).

7. John Demos, "Notes on Life in Plymouth Colony," in *William and Mary Quarterly*, 3rd ser., XXV (1965), 264–86; *idem.*, *A Little Commonwealth*, 192–93; *idem.*, "Old Age in Early New England," Chapter Seven, below.

8. See, for example, findings presented in E. A. Wrigley, *Population and History* (London, 1969), and Thad W. Tate and David L. Ammerman, eds., *The Chesapeake Society in the Seventeenth Century: Essays on Anglo-American Society and Politics* (New York, 1979), *passim*.

9. See Demos, *A Little Commonwealth*, 192–94, and Greven, *Four Generations*, 30–31.

10. See, for example, Henry, *Anciennes Familles Genevoises*, chs. 4–5; E. A. Wrigley, "Family Limitation in Pre-Industrial England," in *Economic History Review*, XIX (1966), 82–109; and Robert V. Wells, "Family Size and Fertility Control in Eighteenth-Century America: A Study of Quaker Families," in *Population Studies*, XXV (1971), 73–82.

11. Lawrence Stone, *The Family, Sex, and Marriage in England, 1500–1800* (New York, 1977), 70; Philippe Ariès, *Centuries of Childhood: A Social History of Family Life*, trans. Robert Baldick (New York, 1962), 38–39.

12. On the extent of this pattern in one early American community, see John Demos, "Families in Colonial Bristol, Rhode Island," in *William and Mary Quarterly*, 3rd ser., XXV (1968), 56–57.

13. Daniel Scott Smith and Michael Hindus, "Premarital Pregnancy in America, 1640–1971: An Overview and Interpretation," in *Journal of Interdisciplinary History*, V (1975), 537–70.

14. This viewpoint informs two older but still important books: Edward Shorter, *The Making of the Modern Family* (New York, 1975) and Fred Weinstein and Gerald M. Platt, *The Wish To Be Free: Society, Psyche, and Value Change* (Berkeley, 1969).

15. See, for example, Greven, *Four Generations*, chs. 3–4.

16. On the pattern in early New England, see Demos, *A Little Commonwealth*, 154–57, and Edmund S. Morgan, *The Puritan Family: Religion and Domestic Relations in Seventeenth-Century New England*, rev. ed. (New York, 1966), 79–86.

17. Daniel Scott Smith, "Parental Power and Marriage Patterns: An Analysis of Historical Trends in Hingham, Massachusetts," in *Journal of Marriage and the Family*, XXXV (1973), 419–28.

18. These passages are excerpted from John Dane, "A Declaration of Remarkable Providences in the Course of My Life," in John Demos, ed., *Remarkable Provinces: The American Culture, 1600–1760* (New York, 1972), 80–88.

19. See, for example, Demos, *A Little Commonwealth*, and Laurel Thatcher Ulrich, *Good Wives: Image and Reality in the Lives of Women in Northern New England* (New York, 1982).

20. Joe L. Dubbert, *A Man's Place: Masculinity in Transition* (Englewood Cliffs, N.J., 1979), ch. 2; E. Anthony Rotundo, "Manhood in America: The Northern Middle Class, 1770–1920," unpublished Ph.D. diss., Brandeis University, 1982; and Rotundo, "Body and Soul: Changing Ideals of American Middle Class Manhood, 1770–1920," in *Journal of Social History*, XVI (1982), 23–38.

21. Nancy F. Cott, *The Bonds of Womanhood: Woman's Sphere in New England, 1780–1835* (New Haven, 1977); Ruth Bloch, "American Feminine Ideals in Transition: The Rise of the Moral Mother, 1785–1815," in *Feminist Studies*, IV (1978), 98–116; Carl Degler, *At Odds: Women and the Family in America From the Revolution to the Present* (New York, 1980), chs. 2–5.

22. See "The Changing Faces of Fatherhood," Chapter Three below.

23. Elizabeth Pleck and Nancy F. Cott, *A Heritage of Her Own* (New York, 1980).

24. Mary Hartman and Lois Banner, *Clio's Consciousness Raised: New Perspectives on the History of Women* (New York, 1974).

25. Gerda Lerner, *A Majority Finds Its Past: Placing Women in History* (New York, 1979).

26. See, for example, Eleanor Flexner, *A Century of Struggle: The Woman's Rights Movement in the United States* (Cambridge, Mass., 1959), and William L. O'Neill, *The Woman Question: Feminism in the United States and England* (Chicago, 1969).

27. See, for example, Elizabeth Anthony Dexter, *Colonial Women of Affairs*, 2nd ed. (Boston, 1931), and Roger Thompson, *Women In Stuart England and America: A Comparative Study* (London, 1974).

28. See Barbara Welter, "The Cult of True Womanhood, 1820–60," in *American Quarterly*, XVIII (1966), 151–74; Gerda Lerner, "The Lady and the Mill Girl: Changes in the Status of Women in the Age of Jackson," in *Mid-Continent American Studies Journal*, X (1969), 5–15; and John Demos, "The American Family in Past Time," in *The American Scholar*, XLIII (1974), 422–46.

29. This revisionist viewpoint is vigorously presented in Mary Beth Norton, "The Evolution of White Women's Experience in Early America," in *American Historical Review*, LXXXIX (1984), 593–619.

30. See Ann Douglas Wood, " 'The Fashionable Diseases': Women's Complaints and Their Treatment in Nineteenth-Century America," in *Journal of Interdisciplinary History*, IV (1973), 25–52; and Carroll Smith-Rosenberg and Charles Rosenberg, "The Female Animal: Medical and Biological Views of Woman and Her Role in Nineteenth-Century America," in *Journal of American History*, LX (1973), 332–56.

31. Ellen K. Rothman, *Hands and Hearts: A History of Courtship in America* (New York, 1984); Degler, *At Odds*, 19–25. The apt term "marriage trauma" was created by Nancy F. Cott; see her *Bonds of Womanhood*, 80.

32. Nancy F. Cott, "Passionlessness: An Interpretation of Victorian Sexual Ideology," in *Signs*, IV (1978), 219–36; and B. J. Barker-Benfield, "The Spermatic Economy: A Nineteenth-Century View of Sexuality," in *Feminist Studies*, I (1972), 45–74.

33. Carroll Smith-Rosenberg, "The Female World of Love and Ritual: Relations Between Women in Nineteenth-Century America," in *Signs*, I (1975), 1–29.

34. See Rotundo, "Manhood in America, 1770–1920," *passim*.

35. Ariès, *Centuries of Childhood*.

36. These and related points are explored at length in "The Rise and Fall of Adolescence," Chapter Five below.

37. On this problem, see Jay Mechling, "Advice to Historians on Advice to Mothers," in *Journal of Social History*, IX (1975), 45–63.

38. Demos, *A Little Commonwealth*, ch. 2; Philip J. Greven, Jr., *The Protestant Temperament: Patterns of Child-Rearing, Religious Experience, and the Self in Early America* (New York, 1977), ch. 9.

39. On nineteenth-century child-rearing, see Bernard Wishy, *The Child and the Republic: The Dawn of Modern American Child Nurture* (Philadelphia, 1968), and Degler, *At Odds*, ch. 5.

40. See John Demos, "Shame and Guilt in Early America," in *Journal of Social History*, forthcoming. The same issue is explored, though not as a matter of historical change, in Greven, *The Protestant Temperament*.

41. Cotton Mather, "Some Special Points Relating to the Education of My Children," in Worthington C. Ford, ed., *The Diary of Cotton Mather*, Massachusetts Historical Society, *Collections*, 7th Ser., VII (1911), 534–37. See also Demos, *A Little Commonwealth*, ch. 9.

42. The fullest, most ambitious of these studies is David Hackett Fischer, *Growing Old in America* (New York, 1977).

43. On such points see Andrew W. Achenbaum, *Old Age in the New Land* (Baltimore, 1978); Carole Haber, *Beyond Sixty-Five: The Dilemma of Old Age in America's Past* (Cambridge, Eng., 1983); and Peter N. Stearns, *Old Age in European Society: The Case of France* (New York, 1976).

44. See Fischer, *Growing Old in America*, chs. 1–2.

45. See "Old Age in Early New England," Chapter Seven below.

46. Shorter, *The Making of the Modern Family* is the most prominent instance of this viewpoint.

47. Benjamin Franklin to Deborah Franklin, 1 May 1771, in Donald M. Scott and Bernard Wishy, eds., *America's Families: A Documentary History* (New York, 1982), 117.

48. This issue is explored at length in "Child Abuse in Context: An Historian's Perspective," Chapter 4 below.

49. Such, at least, is the reaction of "pre-modern" tribespeople in East Africa to modern American handling of infants; Robert and Sarah LeVine, personal communication.

50. Peter N. Stearns and Carol Z. Stearns, "Emotionology: Clarifying the History of Emotions and Emotional Standards," in *American Historical Review*, XC (1985), 813–36.

51. This model is associated, in particular, with the work of Talcott Parsons. See, for example, his "The Social Structure of the Family," in Ruth Anshen, ed., *The Family: Its Functions and Destiny*, rev. ed. (New York, 1959), 241–74; and Parsons and Robert Bales, *Family Socialization and Interaction Process* (Glencoe, Ill., 1955).

52. Demos, *A Little Commonwealth*, 183–85 and *passim*.

53. This viewpoint on contemporary family life can be sampled in Kenneth Keniston and the Carnegie Council on Children, *All Our Children: The American Family Under Pressure* (New York, 1977).

54. See Caroline F. Ware, *The Early New England Cotton Manufacture: A Study in Industrial Beginnings* (New York, 1931). For comparable material from English history, see Michael Anderson, *Family Structure in Nineteenth-Century Lancashire* (Cambridge, Eng., 1971).

55. The newest, most definitive work on this subject is Tamara K. Hareven, *Family Time and Industrial Time: The Relationship Between the Family and Work in a New England Industrial Community* (Cambridge, Eng., 1982).

56. See, for example, James B. Hedges, *The Browns of Providence Plantations*, 2 vols. (Boston, 1952, 1958); Oscar Handlin, *Boston's Immigrants* (Boston, 1941); Mary P. Ryan, *Cradle of the Middle Class: The Family in Oneida County, New York, 1790–1865* (Cambridge, Eng., 1981); and Paul E. Johnson, *A Shopkeeper's Millennium: Society and Revivals in Rochester, New York, 1815–1837* (New York, 1978).

57. David Elkind, *The Hurried Child: Growing Up Too Fast Too Soon* (Reading, Mass., 1981).

CHAPTER II

Images of the Family,
Then and Now

My apprenticeship in family history was strictly pre-professional. Efforts to uncover and organize source materials, to pose questions and frame arguments, to develop a suitable expository style: such were its leading elements. I aimed to add my piece to historical knowledge, to gratify my fellow-historians (teachers and peers), and, not least, to find a niche for myself in one or another college-level History Department.

Family history proved, however, to have an unusually wide orbit of concern. My research drew me repeatedly toward the "social sciences"—toward demography, psychology, anthropology, and sociology. And scholars from those disciplines reciprocated my interest. More and more I found myself operating in an explicitly inter-disciplinary mode.

A little later the orbit swung out in a different direction— toward literature, philosophy, and the "humanities" in general. And from there it was only a short distance further to "policy"— and broadly "public" concerns. Lecture-invitations, conferences, and commission-memberships posed a challenge to the academic historian in me: how to make what I knew meaningful for an audience beyond the bounds of my discipline, indeed of the Academy itself.

The second essay in this volume was one of my first attempts to respond to the challenge. Initially prepared for a conference on family "images," and subsequently presented in a variety of public

This essay was first published in Virginia Tufte and Barbara Myerhoff, eds., *Changing Images of the Family* (New Haven: Yale University Press, 1979), 43–60.

settings, it offers a quick sketch of American family history at large. It invites (at least obliquely) a reader/listener to contrast our contemporary notions of family life with those of our historical predecessors.

In fact, the "images" theme offers an especially good opening to historical contrast. Much early research in family history was directed to demographic questions; and the results landed more on the side of continuity than change. (Thus, for example, "average household size" shows remarkable stability over long periods of time, and even the composition of families has altered less than one might think.) However, the expectations people bring to family life, the rewards they seek from it, the costs they are willing to pay: this is another, far more fluid and variable, story. And "images" cut right to its heart.

Within the past twenty years professional historians in several countries have directed special attention and energy to the study of family life. "Family history" has become for the first time a legitimate branch of scholarly research. The entering wedge of this research was, and remains, demographic; by now we know more about the history of such circumstances as mean household size and median age of marriage than reasonable people may want to know. But investigation has also been moving ahead on other, more "qualitative," tracks: one thinks immediately of recent and challenging studies of the history of childhood, of women, of sexual mores and behavior, and even of domestic architecture.[1]

Meanwhile, the contemporary experience of families has also come under increasingly intense scrutiny. There is a diffuse sense of "crisis" about our domestic arrangements generally—a feeling that the family as we have traditionally known it is under siege, and may even give way entirely. The manifestations of this concern are many and varied—and by now have high visibility. Presidents, legislators, and other government officials regularly invoke the needs of families—and weigh public policy against such needs. One sees as well a new flowering of commissions, task forces, and conferences on this or that aspect of family life—on income supports, on child abuse, on family law, on "parenting."[2]

These two streams of interest have run a similar, but not an intersecting, course. It is arguable that either one might have developed—just about as it actually has developed—had the other never taken life at all. Still, we should perhaps consider the bridges that might usefully be built between them. To put the matter quite simply: what light can a historian throw on the current predicament in family life? For one thing he is tempted right away to strike a soothing note of reassurance. The core structure of the family has evolved and endured over a very long period of Western history, and it is extremely hard to imagine a sudden reversal of so much weighty tradition. Moreover, for at least a century now the American family in particular has been seen as beleaguered, endangered, and possibly on the verge of extinction. The sense of crisis is hardly new; with some allowance for periodic ebb and flow, it seems an inescapable undercurrent of our modern life and consciousness.

Is this, in fact, reassuring? And does such reassurance help, in any substantial way? Somehow, historians must try to do better.

These considerations will serve to frame (and perhaps to excuse) certain features of the present essay. The scale of discussion will be very large indeed—nothing less than the entire sweep of American history. The substance will be somewhat more modest, with the focus held for the most part on "images" of the family (not behavior); but even this encompasses a broad, and highly variable, territory. The tone will be, at least occasionally, judgmental and partisan; even the most "objective" scholar need not wholly suppress his own sense of gains derived, and prices paid, from the central perceptions of domestic life in the past.

More specifically, the essay proposes a three-part model of family history as a way of periodizing the field. In doing so it necessarily lumps together many highly variegated bits of research; and, obviously, other histories—other historians—might fashion quite a different set of lumps. In emphasizing images of family life, it accepts an implicit social and economical bias, for such images have been created largely by people of Anglo-Saxon origin in the more comfortable layers of our social system—in short, by middle-class WASPs. Nonetheless this same middle class has, in our country, traditionally played a style-setting role—even for those who might seem to espouse alternative ways and traditions. Immigrants, blacks, workers of all sorts, are thus profoundly implicated here. The nature of these connections is highly complex— one might well say, ambivalent. But the larger point is that Americans of every color, every creed, and every economic position have been drawn toward the cultural middle. And embedded just there are the very images that form the subject of this discussion.

I

Consider, as the first part of our sequential model, the "colonial" and "early national" phases of American history—a period of time lasting into the nineteenth century. It should be recognized, incidentally, that this inquiry does not lend itself to precise chronological markings; the end of one stage and the start of the next are so fully merged that one might well think in terms of a transitional process. And if, for the first stage, one assigns a terminal date of 1820, this is only to indicate a midpoint in the transition. The precursors of change were visible in some quarters as early as the Revolutionary era, and the process was still working itself out during the time of the Civil War.

Now comes a rather perverse twist. When a scholar seeks to approach the colonial American family, one thing he notices immediately is that the "image" itself seems rather thinly sketched. In short, people of that rather distant time and culture did not have a particularly self-conscious orientation to family life; their ideas, their attitudes in this connection, were far simpler than would ever be the case for later generations of Americans. Family life was something they took largely for granted. It was no doubt a central part of their experience, but not in such a way as to require special attention. This does not mean that they lacked ideas of what a "good family" should be and do—or, for that matter, a "bad" one—just that such notions carried a rather low charge in comparison with other forms of social concern.[3]

In any event, there are some points that one can deduce about their orientation to family life, and a few that can be pulled directly out of the documentary record they have left to us. Here is a particularly resonant statement, taken from an essay by a "Puritan" preacher in the early seventeenth century:

> A family is a little church, and a little commonwealth, at least a lively representation thereof, whereby trial may be made of such as are fit for any place of authority, or of subjection, in church or commonwealth. Or rather, it is as a school wherein the first principles and grounds of government are learned; whereby men are fitted to greater matters in church and commonwealth.[4]

Two aspects of this description seem especially important. First, the family and the wider community are joined in a relation of profound reciprocity; one might almost say they are continuous with one another. (This is, incidentally, a general premodern pattern—in no sense specific to American life and conditions—which was analyzed first and most incisively by Philippe Ariès, in his path-breaking study of twenty years ago published in English under the title *Centuries of Childhood*.[5]) To put the matter in another way: individual families are the building blocks out of which the larger units of social organization are fashioned. Families and churches, families and governments, belong to the same world of experience. Individual people move back and forth between these settings with little effort or sense of difficulty.

The membership of these families was not fundamentally different from the pattern of our own day: a man and a woman joined in marriage, and their natural-born children. The basic unit was therefore a "nuclear" one, contrary to a good deal of sociological theory about premodern times. However, non-kin could, and did, join this unit—orphans, apprentices, hired laborers, and a variety of children "bound out" for a time in conditions of fosterage. Usually designated by the general term "servants," such persons lived as regular members of many colonial households; and if they were young, the "master" and "mistress" served *in loco parentis*. Occasionally, convicts and indigent people were directed by local authorities to reside in particular families. Here the master's role was to provide care, restraint, and even a measure of rehabilitation for those involved; they, in turn, gave him their service. Thus did the needs of the individual householders intersect the requirements of the larger community.[6]

But it was not simply that the family and the community ran together at so many points; the one was, in the words of the preacher, "a lively representation" of the other. Their structure, their guiding values, their inner purposes, were essentially the same. Indeed the family was a community in its own right, a unit of shared experience for all its individual members. It was, first and foremost, a community of work—in ways hard for us even to imagine today. Young and old, male and female, labored together to produce the subsistence on which the whole group depended. For long periods they worked literally in each other's presence—if not necessarily at the same tasks. In other ways as well the family lived and functioned as a unit. Most leisure-time activities (which consisted largely of visiting with friends, relatives, and neighbors) were framed in a family context, as were education, health care, and some elements of religious worship.

Religion, however, brings new considerations into view; in effect, it conjures up another, quite different community. Clarity on this point requires a brief detour into the esoteric particulars of church organization. Consider the typical seating plan of colonial churches (at least in New England).[7] (1) Men and women were separated on opposite sides of a central aisle. (2) Within these sex-typed enclaves individual communicants were assigned places in accordance with certain "status" criteria. (In general, the oldest, wealthiest, and most prominent citizens sat at the front.) (3) Children were relegated to still another section of the church (usually in the back, sometimes in an upstairs gallery). Now these arrangements have an important bearing on the present inquiry. Presumably, the members of a typical family started out for church together, but upon reaching their destination they broke apart and went off in different directions. The church had its own mode of organization, vividly reflected in its spatial aspect; preacher in the pulpit, elders sitting just below, regular parishioners carefully distributed on the basis of sex, age, and social position. Family relationships were effectively discounted, or at very least submerged, in this particular context. In sum, the family community and the religious community were fundamentally distinct— though formed from the same pool of individuals. Much later (in most cases, the early nineteenth century), Protestant congregations went over to the pattern that still prevails today, with seating inside the church arranged on a family basis. And the change had a profound symbolic significance. No longer was the family simply one form of community among others; from thenceforth it constituted a special group, whose boundaries would be firmly declared in all imaginable circumstances.

There is one more vital aspect of colonial family life which deserves at least to be mentioned. Since the functions of the household and the wider society were so substantially interconnected, the latter might reasonably intervene when the former experienced difficulty. Magistrates and local officials would thus compel a married couple "to live more peaceably together" or to alter and upgrade the "governance" of their children. This, too, is the context of the famous "stubborn child" laws of early New England, which prescribed the death penalty for persistent disobedience to parents. Such extreme sanctions were never actually invoked, but the statutes remained on the books as a mark of society's interest in orderly domestic relations.[8]

II

As noted earlier, it is hard to say just how and when this colonial pattern began to break down; but by the early decades of the nineteenth century at least some American families were launched on a new course, within a very different framework of experience. For the most part these were urban families, and distinctly middle-class; and while they did not yet constitute anything like a majority position in the country at large, they pointed the way to the future.[9]

Here, for the first time, American family life acquired an extremely sharp "image"—in the sense of becoming something thought about in highly self-conscious ways, written about at great length and by many hands, and worried about in relation to a host of internal and external stress-points. Among other things, there was a new sense that the family had a history of its own—that it was not fixed and unchanging for all time. And when some observers, especially "conservative" ones, pondered the direction of this history, they reached an unsettling conclusion: the family, they believed, was set on a course of decline and decay. From a stable and virtuous condition in former times, it had gradually passed into a "crisis" phase. After mid-century, popular literature on domestic life poured out a long litany of complaints: divorce and desertion were increasing; child-rearing had become too casual and permissive; authority was generally disrupted; the family no longer did things together; women were more and more restless in their role as homemakers.[10] Do these complaints have a somewhat familiar ring even now? In fact, it is from this period, more than one hundred years ago, that one of our most enduring images of the family derives—what might be called the image, or the myth, of the family's golden past. Many around us appear still to believe that there is some ideal state of domestic life which we have tragically lost. The consequences of such belief are profound—since it implies that our individual efforts and our public policies in regard to family life should have a "restorative" character. But this is another topic, best left to a different setting.

How, then, shall we further characterize the nineteenth-century image of family-life? One point is immediately striking. The family—far from joining and complementing other social networks, as in the earlier period—seemed to stand increasingly apart. Indeed its position vis-à-vis society at large had been very nearly reversed, so as to become a kind of adversary relation.

The brave new world of nineteenth-century America was, in some respects, a dangerous world—or so many people felt. The new egalitarian spirit, the sense of openness, the opportunities for material gain, the cult of the "self-made man": all this was new, invigorating, and liberating at one level—but it also conveyed a deep threat to traditional values and precepts. In order to seize the main chance and get ahead in the ongoing struggle for "success," a man had to summon energies and take initatives that would at the very least exhaust him and might involve him in terrible compromises. At the same time he would need to retain some place of rest and refreshment—some emblem of the personal and moral regime that he was otherwise leaving behind.[11]

Within this matrix of ideas the family was sharply redefined. Henceforth the life of the individual home, on the one hand, and the wider society, on the other, represented for many Americans entirely different spheres. ("Spheres" was indeed the term they most frequently used in conceptualizing their varied experiences.) The two were separated by a sharply delineated frontier; different strategies and values were looked for on either side.

Some of the new "home" values have been mentioned already, but it is necessary to investigate them more fully. Home—and the word itself became highly sentimentalized—was pictured as a bastion of peace, of orderliness, of unwavering devotion to people and principles beyond the self. Here the woman of the family, and the children, would pass most of their hours and days—safe from the grinding pressures and dark temptations of the world at large; here, too, the man of the family would retreat periodically for repose, renewal, and inner fortification against the dangers he encountered elsewhere.[12]

Pulling these various themes together, one can reasonably conclude that the crucial function of the family had now become a protective one. And two kinds of protection were implied here: protection of the ways and values of an older America that was fast disappearing, and protection also for the individual people who were caught up in the middle of unprecedented change. If the notion of "the family as community" serves to summarize the colonial part of this story, perhaps for the nineteenth century, the appropriate image is "the family as refuge." Two short passages, chosen from the voluminous domestic literature of the period, will convey the underlying melody:

> From the corroding cares of business, from the hard toil and frequent disappointments of the day, men retreat to the bo-soms of their families, and there, in the midst of that sweet society of wife and children and friends, receive a rich reward for their industry. . . . The feeling that here, in one little spot, his best enjoyments are concentrated . . . gives a wholesome tendency to [a man's] thoughts, and is like the healing oil poured upon the wounds and bruises of the spirit.

> We go forth into the world, amidst the scenes of business and of pleasure; we mix with the gay and the thoughtless, we join the busy crowd, and the heart is sensible to a desolation of feeling; we behold every principle of justice and of honor, disregarded, and the delicacy of our moral sense is wounded; we see the general good sacrificed to the advancement of per-sonal interest; and we turn from such scenes with a painful sensation, almost believing that virtue has deserted the abodes of men; again, we look to the sanctuary of home; there sym-pathy, honor, virtue are assembled; there the eye may kindle with intelligence, and receive an answering glance; there dis-interested love is ready to sacrifice everything at the altar of affection.[13]

This imagery had, in fact, particular features which deserve careful notice. For one thing, it embraced the idea of highly differ-entiated roles and statuses within the family—for the various indi-vidual family members. The husband-father undertook an exclu-sive responsibility for productive labor. He did this in one or another setting well removed from the home-hearth, in offices, factories, shops, or wherever. So it was that family life was wrenched apart from the world of work—a veritable sea-change in social history. Meanwhile, the wife-mother was expected to con-fine herself to domestic activities; increasingly idealized in the figure of the "True Woman," she became the centerpiece in the developing cult of Home. Intrinsically superior (from a moral standpoint) to her male partner, the True Woman preserved Home as a safe, secure, and altogether "pure" environment.[14] The chil-dren of this marital pair were set off as distinctive creatures in their own right. Home life, from their point of view, was a sequence of preparation in which they armored themselves for the challenges and difficulties of the years ahead.[15] The children, after all, carried the hopes of the family into the future; their lives later on would reward, or betray, the sacrifices of their parents. Taken altogether, and compared with the earlier period, these notions conveyed the sense of a family carefully differentiated as to individual task and function, but unalterably united as to overall goals and morale. Like other institutions in the "Machine Age," the family was now seen as a system of highly calibrated, interlocking parts.

It is clear enough that such a system conformed to various practical needs and circumstances in the lives of many Americans—the adaptation to urban life, the changing requirements of the workplace, the gathering momentum of technology. But it must have answered to certain emotional needs as well. In particular, the cult of Home helped people to release the full range of aggressive energies so essential to the growth and development of the country—helped them, that is, to still anxiety and to ward off guilt about their own contributions to change. At the same time there were costs and difficulties that one cannot fail to see. The demands inherent in each of the freshly articulated family roles were sometimes literally overwhelming.

The husband-father, for example, was not just the breadwinner for the entire family; he was also its sole representative in the world at large. His "success" or "failure"—terms which had now obtained a highly personal significance—would reflect directly on the other members of the household. And this was a grievously heavy burden to carry. For anyone who found it too heavy, for anyone who stumbled and fell while striving to scale the heights of success, there was a bitter legacy of self-reproach—not to mention the implicit or explicit reproaches of other family members whose fate was tied to his own.

Meanwhile, the lady of the house experienced another set of pressures—different, but no less taxing. The conventions of domestic life had thrown up a model of the "perfect home"—so tranquil, so cheerful, so pure, as to constitute an almost impossible standard. And it was the exclusive responsibility of the wife to try to meet this standard. Moreover, her behavior must in all circumstances exemplify the selflessness of the True Woman. Her function was effectively defined as one of service and giving to others; she could not express needs or interests of her own. This suppression of self exacted a crushing toll from many nineteenth-century women. Few complained outright, though modern feminism dates directly from this era. But there were other, less direct forms of complaint—the neurasthenias, the hysterias, indeed a legion of "women's diseases," which allowed their victims to opt out of the prescribed system.[16]

The system also imposed new difficulties on the younger members of the household. In the traditional culture of colonial America the process of growth from child to adult had been relatively smooth and seamless. The young were gradually raised, by a sequence of short steps, from subordinate positions within their families of origin to independent status in the community at large.[17] In the nineteenth century, by contrast, maturation became disjunctive and problematic. As the condition of childhood was ever more sharply articulated, so the transition to adulthood became longer, lonelier, more painful.[18] And there was also another kind of transition to negotiate. For those who absorbed the imagery of Home the moment of leaving was charged with extraordinary tension. To cross the sacred threshold from inside to outside was to risk unspeakable dangers. The nostalgia, the worries, the guilt which attended such crossings are threaded through an enormous mass of domestic fiction from the period. Marriage itself was experienced as the sudden exchange of one family for another—with a little of the flavor of a blind leap.[19]

In sum, the "ideal family" of the nineteenth century comprised a tightly closed circle of reciprocal obligations. And the entire system was infused with a strain of dire urgency. If the family did not function in the expected ways, there were no other institutions to back it up. If one family member fell short of prescribed ways and standards, all the others were placed in jeopardy. There is a short story by T. S. Arthur—an immensely popular author during the middle of the century—which makes this point very clearly.[20] A young couple marry and set up housekeeping. The husband is an aspiring businessman, with every prospect of "success." His wife shares his ambitions and means to become an effective "helpmeet"; however, her management of the household is marred by a certain inefficiency. The husband regularly returns from his office for lunch (an interesting vestige of premodern work rhythms), but soon a problem develops. The wife cannot hold to a firm schedule in preparing these meals, and often her husband is kept waiting. Earnest conversations and repeated vows of improvement bring no real change. Finally, one particular delay causes the husband to miss a crucial appointment—and the consequences for his business are devastating.

Domestic fiction played out similar themes in the relation of parents and children. Only the most careful and moral "rearing" would bring the young out safe in later life; anything less might imperil their destiny irrevocably. Conversely, the well-being of parents depended in large measure on their offspring. If the latter, having grown to adulthood, were to stray from the paths of virtue, the old folks might feel so "heartbroken" that they would sicken and die.[21] Here the stakes of domestic bonding attained an aspect of life-threatening finality.

III

To some degree the image of "the family as refuge" remains with us today. Many people still look to home life for buffering, or at least for relief, against the demands and pressures of society at large. There is a continuing sense of inside-outside, and an idea of domestic inviolacy well expressed by the cliche that "each man's home is his castle."

And yet for some time the tide has been running in another direction. It seems reasonable, therefore, to posit a third stage in family history, while acknowledging that the second has not yet exhausted itself. To identify the new trend precisely is no easy task, and the argument of the following pages must be regarded as provisional at best. A scholar is *perforce* reduced to the role of an everyday observer—indeed a participant-observer (!)—of contemporary family life.

Domestic imagery can be expected to reflect changes in the social context of experience, and for millions of Americans that context is now a different one. The twentieth century has gathered up a host of "modernizing" forces; one overused but still helpful phrase for describing the process is "the rise of mass society." Truly, we do take significant parts of life in the mass—as workers (mostly for large organizations), as consumers (mostly of highly standardized services and products), as citizens (mostly under a "big government"). For many Americans this situation has brought a measure of security and comfort unprecedented in previous generations. Jobs are less liable to disruption, income is steadier, health care is somewhat more regular, and so on. And yet these gains have been purchased at a cost. Comfortable as many of us are, we have a sense of flatness, even of emptiness, about large sectors of our experience. Increasingly we feel that we are not masters of our own fate, that our individual goals and deeds count for nothing when weighed in such a large aggregate. We cannot, in short, make much of a difference in our own lives. Thus ever larger numbers of us do not bother to vote (what difference is made by one ballot more or less?), and we are disinclined to protest inferior products, inefficient services, or even blatant injustice when such things directly touch our lives. "Apathy" is the currently fashionable word to describe our social climate—and it does seem to hit the mark.

For a shorthand contrast between this situation and the social climate of a century ago one need only consult the favorite period metaphors. History has moved, it seems, from the "jungle" of the nineteenth century to the "rat race" (or the "grind") of our own day. This progression expresses clearly a lessened sense of threat—and also a growing feeling of monotony and meaninglessness.

The implications for family life—specifically, for images and expectations of family life—are profound. As the threat is tempered, the wish for protection, for armoring, wanes. Or rather it shades gradually into something else. Home is less a bunker amidst the battle than a place of "rest and recuperation" (pursuing the military analogy). According to this standard, families should provide the interest, the excitement, the stimulation missing from other sectors of our experience. If we feel that "we aren't going anywhere" in our work, we may load our personal lives—especially our family lives—with powerful compensatory needs. We wish to "grow" in special ways through our relations with family partners; a familiar complaint in counseling centers nowadays is the sense of blocked opportunities for growth. We want our spouses, our lovers, even our children, to help us feel alive and invigorated—to brighten a social landscape that otherwise seems unrelievedly gray. Again, some contrasts with the earlier setting may be helpful. Then Home was to be a place of quiet, of repose. Now it must generate some excitement. Then the True Woman served as the appointed guardian of domestic values; as such she was "pure," steady, in all ways self-effacing. Now there is the figure of the "Total Woman"—who, to be sure, keeps an orderly house and seeks consistently to help her man, but who is also sensual and assertive within limits.[22]

Indeed an entire spectrum of roles and responsibilities within the family is increasingly in question. No longer can we automatically accept that principle of differentiation which, in the nineteenth century, assigned to each household member a "sphere" deemed appropriate to his or her age and gender. Some families now advocate an opposite principle, which exalts the diffusion and mixing of roles. Mother must do her share of the "breadwinning," Father must do his share of the household chores, and so on. Much of this, of course, comes directly from the "women's movement," and involves a long-overdue effort to right the balance of opportunity between the sexes. However, that is not the whole story. If Father is urged nowadays to help with the children or wash the dishes or take care of the laundry, this is not just in order to lighten the burdens that have traditionally fallen on Mother. There is also a feeling that such activities are good for him. Somehow his sensibilities will be expanded and his personal growth advanced—just as Mother expands and grows through her work outside the home. As a further benefit of these rearrangments the couple discovers new byways of marital communication. Since they share so much more, they understand each other better; and their relation to one another is "deepened" accordingly. Even children are invited to join in this celebration of openness and reciprocity. The parents believe that they must listen carefully and at all times to their children, even that they can learn from their children—ideas which would have seemed quite preposterous just a few generations ago.

If all goes well—if reality meets expectation and conforms to image—Home becomes a bubbling kettle of lively, and mutually enhancing, activity. But, alas, all does not invariably go well; so we also have, for the first time in American history, a negative image—an "anti-image"—of the family. Seen from this viewpoint, domestic relationships look dangerously like an encumbrance, if not a form of bondage, inhibiting the quest for a full experience of self. Monogamous marriage is liable to become boring and stultifying; in other things, after all, variety is "the spice of life." Moreover, responsibility for children only compounds the problem. The needs and requirements of the young are so pressing, so constant, as to leave little space for adults who must attend to them. "Spice" and "space": these are, in fact, the qualities for which we yearn most especially. And the family severely limits our access to either one.

These contrasting notions of family-life—image and anti-image—appear at some level to converge. In fact, they are opposite faces of the same coin. Each affirms the primacy of family experience in relation to larger goals of personal growth and self-fulfillment. The difference lies in the effects imputed to such experience—in the first case, a beneficial effect, in the second, an adverse one. Both images assume a deep threat of inward stagnation, implicit in the "rat race" which surrounds us all.

There are, of course, other ways to fight the rat race. Encounter-groups, assertiveness-training, consciousness-raising of all kinds should certainly be mentioned here—as well as the vicarious excitements that come, for example, from celebrity-watching. Presumably these activities are not antithetical to good, growing family life; on the contrary, their effects should be complementary and enhancing. Indeed the most suitable caption for this third (current) stage of family history is: "the family as encounter group." For the central values attaching to domestic experience nowadays—at least by the reading presented here—are those which underscore significant personal encounters.

IV

In closing, it seems necessary to reemphasize the distortion introduced by any discrete model of family history. The problem is not simply one of arbitrary chronological boundaries; there is also a risk of failing to see the cumulative element in all historical process. "Stage three," as previously noted, retains a good part of "stage two"—and even some traces of "stage one." (We continue, after all, to see individual families as the "building blocks" of the nation as a whole.) Thus our present arrangements are best construed as a complex and heavily layered precipitate of our entire social history.

Two points about this history deserve some final underscoring. In both the second and the third of our major stages the family has been loaded with the most urgent of human needs and responsibilities. Indeed one might well say overloaded. In each case the prevalent imagery conjures up a compensatory function: the family must supply what is vitally needed, but missing, in social arrangements generally. It must protect its individual constituents against imminent and mortal danger, or it must fill a void of meaninglessness. To put the matter in another way: the family is not experienced in its own right and on its own terms, but in relation to outside circumstances and pressures. It is for this reason, presumably, that we have become so extraordinarily self-conscious about family-life—and, more, have broached it for so long from an attitude of crisis.

There is a concomitant of this attitude which also has deep historical roots. Briefly, we have isolated family life as the primary setting—if not, in fact, the only one—for caring relations between people. The nineteenth-century images made an especially powerful contribution here: each family would look after its own—and, for the rest, may the best man win. Relationships formed within the world of work—which meant, for a long time, relationships between men—would not have an emotional dimension. Nineteenth-century women seem, on the evidence of very recent scholarship, to have maintained "networks" of affection which did cross family boundaries, but even this pattern recedes as we follow the trail toward the present.

Much of the same viewpoint has survived into our own time, and it underlies certain continuing tensions in our national experience. The United States stands almost alone among Western industrialized countries in having no coherent "family policy." More particularly, our inherited habits and values—our constricted capacity for extra-familial caring—partly explain public indifference to the blighted conditions in which many families even now are obliged to live. The results are especially tragic as they affect children, and they leave us with a terrible paradox. In this allegedly most child-centered of nations, we find it hard to care very much or very consistently about *other people's children*. A historian may think that he understands such a predicament; he does not, however, know how to change it.

NOTES

1. See the discussion (and references) in Chapter One, above.

2. Several recent publications exemplify the trend: e.g. Kenneth Keniston and the Carnegie Council on Children, *All Our Children: The American Family Under Pressure* (New York, 1977); Assembly of Behavioral and Social Sciences, National Research Council, *Toward a National Policy for Children and Families* (Washington, D.C.: National Academy of Sciences, 1976). See also the various publications of the Family Impact Seminar Series (Temple University Press) and the Committee on Child Development Research and Public Policy (National Research Council, Washington, D.C.).

3. On the colonial family, see, for example, John Demos, *A Little Commonwealth: Family Life in Plymouth Colony* (New York, 1970), Edmund S. Morgan, *The Puritan Family*, rev. ed. (New York, 1966), and Daniel Blake Smith, *Inside the Great House: Planter Family-Life in Eighteenth-Century Chesapeake Society* (Ithaca, N.Y., 1980).

4. William Gouge, *Of Domesticall Duties* (London, 1622).

5. See Philippe Ariès, *Centuries of Childhood: A Social History of Family Life*, trans. Robert Baldick (New York, 1962).

6. See Morgan, *Puritan Family*, ch. 4.

7. On the practice of "seating the meetinghouse," see Robert J. Dinkin, "Provincial Massachusetts: A Deferential or a Democratic Society" (Ph.D. diss., Columbia University, 1968), and Ola Winslow, *Meetinghouse Hill, 1630–1783* (New York, 1952).

8. See Edwin Powers, *Crime and Punishment in Early Massachusetts, 1620–1692: A Documentary History* (Boston, 1966), 268, 283ff.

9. See Kirk Jeffrey, "The Family as Utopian Retreat from the City," *Soundings* 55 (1972): 21–41; and Barbara Laslett, "The Family as a Public and Private Institution: An Historical Perspective," *Journal of Marriage and the Family*, 35 (1973):480–92.

10. Examples of this viewpoint may be found in William A. Alcott, *The Young Wife* (Boston, 1837), William Thayer, *Hints for the Household* (Boston, 1853), and Artemas B. Muzzey, *The Fireside* (Boston, 1854).

11. See Jeffrey, "Family as Utopian Retreat." Also Barbara Welter, "The Cult of True Womanhood: 1820–1860," *American Quarterly* 18 (1966):151–74.

12. An immensely popular expression of this viewpoint was the novel by Catharine Maria Sedgwick, *Home* (Boston, 1854).

13. Quoted in Jeffrey, "Family as Utopian Retreat."

14. See Welter, "Cult of True Womanhood."

15. See Bernard Wishy, *The Child and the Republic: The Dawn of Modern American Child Nurture* (Philadelphia, 1972).

16. See Ann Douglas Wood, "'The Fashionable Diseases': Women's Complaints and Their Treatment in Nineteenth-Century America," *Journal of Interdisciplinary History* 4 (1973): 25–52; and Carroll Smith-Rosenberg, "The Hysterical Woman: Some Reflections on Sex-Roles and Role Conflict in Nineteenth-Century America," in *Social Research*, 39 (1972), 652–78.

17. Demos, *A Little Commonwealth*, ch. 10.

18. Joseph Kett, *Rites of Passage: Adolescence in America, 1790 to the Present* (New York, 1977).

19. See, for example, the marriages described in Sedgwick, *Home*. This theme is explored in Ellen K. Rothman, *Hands and Hearts: A History of Courtship in America* (New York, 1984).

20. T. S. Arthur, "Sweethearts and Wives," in *The Root of Bitterness*, ed. Nancy F. Cott (New York, 1972).

21. The plot line in Sedgwick's novel is a case in point.

22. Marabel Morgan, *The Total Woman* (Old Tappan, N.J., 1975).

CHAPTER III

The Changing Faces of Fatherhood

In family history, as on sinking ships, the priority has been "women and children first." The mother-child pair holds the focus, whenever inquiry turns to key, internal themes and questions. There are many reasons for this, not least the fact that women and children do seem to occupy central positions in families—do so now, and have for at least the past century. In life as in scholarship, then, we note an absence—*the missing man. Yet we should not assume that men have always been missing. Indeed, it could well be argued that men's experience of domestic life has changed more deeply than that of all the other players combined.*

Curiously, these considerations did not dawn on me until I had been through many long years of research. And they might not have dawned on me yet—but for an invitation proferred by three psychiatrists. In psychiatry (and psychology, too) recent years have brought a marked growth of interest in fatherhood. What is a "good father?" How are fathers different from mothers? Where does fathering fit, or not fit, with other parts of a man's experience? Such questions, at once intrinsic to the human condition and deeply culture-bound, have raised up a flurry of activity within the so-called "helping professions."

Inevitably, some of this activity has involved publishing. And so it was that the aforementioned psychiatrists found me. They were planning a volume of "clinical and research contributions" on contemporary fatherhood, and they wished to include an "his-

This essay was first published in Stanley H. Cath, Alan R. Gurwitt, and John Munder Ross, eds., *Father and Child: Developmental and Clinical Perspectives* (Boston, 1982), 425–45.

41

torical overview." How could they obtain such a piece? What scholars were already at work on fatherhood history? The answer was none—at least none that I knew—but without much prodding I rose to the bait.

Looking back on this experience, and looking ahead to what is sure to follow, one can see a new current of study slowly taking form. Fatherhood history makes a single part of a much larger agenda—"men's history," as some have begun to call it. Of course, men have always played a prominent part in the history that scholars have chosen to tell. But this has been, till very recently, a history of public life and official behavior (where the leading figures are almost exclusively male). The rise, within the past dozen years or so, of the subfield "women's history" has necessarily spotlighted private life and personal behavior. And it has helped to show us how little we know about any aspect of past private life.

There is a curious irony here. Women's history was initiated, in part, as an effort to right an imbalance of emphasis—to give to women at least a small portion of the attention hitherto lavished on men. Yet it has begun to stimulate additional study of men themselves.

Fatherhood has a very long history, but virtually no historians.* Its only invariant aspect is the biological one: all else is fluid and changing. No two individuals father in precisely the same way; similarly, no two cultures, or historical epochs, support identical styles of fathering. Of course, there is always overlap—shared elements of purpose, of practice, and of emotional style—especially in adjacent settings. Moreover, where the settings are historically adjacent, we may reasonably expect continuities as well as contrasts. Still, the point remains that fatherhood, no less than other parts of human experience, bends to the passage of time.

But if this serves to define history, it does not automatically give rise to scholarship. Historians choose from a huge field of topical possibility, and they have not as yet chosen fatherhood.

*I would like to acknowledge a profound debt to E. Anthony Rotundo, formerly a member of the doctoral program in the History of American Civilization at Brandeis University, and also to Rachel Cramer, a recent alumna of Brandeis. Their research underlies much of what I have attempted in this essay.

Public events and collective experience have traditionally claimed their interest—history as "past politics." Indeed, for earlier generations of historians fatherhood would scarcely have been legitimate—and perhaps not even conceivable—as a center of serious investigation. Even with the recent flowering of a so-called "new social history," fatherhood remains out of view. There are no focused treatments of the subject in print—no monographs, articles, or compilations of data, let alone synthetic overviews. The historical study of fatherhood is waiting to be born.

Fortunately, the wait will not last much longer if current trends and portents hold firm. The new social history carries within it the seeds of gestation, and the larger scholarly environment looks more and more favorable. Interest in personal, even private, experience has been clearly legitimized. Recent work in family history, women's history, and childhood history has raised implicit questions about fatherhood, and has thrown up promising material for answering such questions. Points of intersection with each of these lively subspecialities are obvious—and advantageous. Problems and *lacunae* will be more an inducement than a hindrance to study. In fact, younger scholars are already busy around the edges of the field, and the results of their work will be available within a few short years.[1]

The present essay is, then, a prediction of birth—although not the event itself. It anticipates progress, raises questions, flags problems, and tries finally to fashion an outline for the history of fatherhood in America. Hence it qualifies as what one scholar has called "hypothetical history"—that is, an informed guess, in advance of the requisite base in monographic research. This is the way things *probably* were—no more, and perhaps less.

I

A vast gulf of change separates early American fathers from their counterparts today. The differences embrace underlying goals and values; prescribed methods and styles of practice; the shape and quality of personal interaction; and the larger configuration of domestic life. Consider the following entry in the diary of a New England clergyman near the end of the seventeenth century:

> I took my little daughter Katy into my study and there I told
> my child that I am to die shortly, and she must, when I am
> dead, remember everything that I said unto her. I set before her

> the sinful and woeful condition of her nature, and I charged
> her to pray in secret places every day without ceasing that God
> for the sake of Jesus Christ would give her a new heart. . . . I
> gave her to understand that when I am taken from her she
> must look to meet with more humbling afflictions than she
> does now [when] she has a careful and a tender father to
> provide for her.[2]

This vignette of "tender fatherhood" startles us: to invoke for a
young child the spectre of parental death seems thoughtless and
manipulative, if not patently cruel. In fact, the parent in question
was young and in good health; his actual death lay decades in the
future. But viewed in context, his admonitions seem less peculiar
and unsettling, and may even make a certain sense. Death was an
active presence in the lives of early Americans, old and young; and
a father who did not prepare his child for such possibilities might
well be considered negligent. Note, too, that discussion of death
was directly linked to moral and religious instruction; the end in
view was the child's improvement in "grace," and even her ultimate "salvation." Whether tender or not, this father seems concerned, involved, active for the welfare of his children. And other
evidence of his family experience points strongly in the same
direction.

In fact, the picture sketched here fits nicely with a large corpus
of prescriptive statements from the period. When ministers and
others in positions of leadership wrote about fatherhood, they
emphasized a broad range of tasks and responsibilities. Father
must be centrally concerned in the moral and religious education
of the young. He must impart the rudiments of reading and writing—to the extent, at least, of his own literacy (*father as pedagogue*). He would play a key role, for both sons and daughters, in
courtship and marriage-making: approving (or disapproving) a
proposed match, and allotting "portions" of family property to
secure a couple's future (*father as benefactor*). The power to deny
or delay such benefaction was also important; thus, in some instances, did aging men continue to hold even adult sons on a very
short leash (*father as controller*). By contrast, mother's part in all
this was rarely, and barely, mentioned. Some authorities felt
obliged to caution that she should not be "exempted"—in implicit
but clear recognition of her lesser role.[3]

There is the further point that most such prescription was
addressed to men (or to parents generally, but with the operative

pronoun "he"). And if we ask why this was so, we must turn to larger questions of gender. Beliefs about maleness and femaleness are also historically variable, as we ourselves can testify from recent social experience. Our forebears of two and three centuries ago maintained some characteristic attitudes toward gender considerably at variance with our own. Men, they believed, must "overrule" women, in domestic affairs no less than other spheres of activity. For men had received from their Maker a superior endowment of "reason." Both sexes were liable to be misled by the "passions" and the "affections," but women were *more* liable because their rational powers were so weak. The biblical account of creation, and of Eve's temptation in the Garden, made this point clearly enough, and there were numerous personal examples from everyday life. A well-known passage in the journal of John Winthrop (first governor of Massachusetts) describes an unfortunate lady thought to have gone insane "by giving herself largely to reading and writing." To woman's intellectual inferiority was added her special *moral* vulnerability. It was no accident that most witches turned out to be female.[4]

This complex of belief and practice carried important implications for parenting. Children came into the world inherently "stained" with sin; moreover, their "passions" were immediately powerful, their "intellectuals" miserably underdeveloped. The steady hand of fathers was necessary to restrain the former, while encouraging and molding the latter (*father as moral overseer*). Unfortunately, the influence of mothers frequently ran the opposite way. "Indulgence," "excessive fondness," a tendency to spoil or "cocker" the children: such were the common maternal failings. Carried to and fro by their inordinate affections, and lacking the "compass" of sound reason, women could hardly provide the vigilant supervision that all children needed. Too, men were better positioned (than women) to understand the young; as one authority put it, "Father ordinarily has the most share in procuring, and most sense in perceiving, the wisdom of his children" (*father as psychologist*). Finally, fathers provided the best models of good character and right behavior (*father as example*). In all these ways and more would men predominate as parents.[5]

To be sure, the pattern was modified by particular circumstances—for example, the age and gender of the children. We can safely assume that infants were largely in the day-to-day care of mothers, for breast milk was their chief source of nourishment, and this alone must have dictated a substantial maternal presence.

Moreover, daughters remained (in general) closer to mothers than did sons; their common gender, and a shared round of household tasks, forged emotional bonds of lasting strength. Still, these are only modifications, not refutations, of the basic rule. Once infants were past the age of breast-feeding, their fathers came strongly into view; and girl children, no less than boys, required moral supervision from a man. It was chiefly for this reason that the common law affirmed overall rights of child custody to the father in cases of marital separation.[6]

The closeness of fathers and sons can be studied, and tested, in several ways. Patterns of correspondence are notably revealing. Teenage boys serving apprenticeships, as well as young men officially on their own, would maintain contact with their families of origin chiefly through letters to fathers. Often they would ask to be "remembered" also to their mothers—but in terms that seem (by our lights) formal, or downright perfunctory. Of letters written directly to mothers there were very few. One man whose father had just died wrote home to a brother and included the following message to their mother: "I sincerely condole with . . . [her] on the loss of her husband; please tender my duty to her." Here, in a single sentence, and especially in the reference to "her husband," is palpable expression of the distance thought appropriate to the relation of mothers and sons.[7]

It appears, too, that fathers—not mothers—were particularly identified with the prospects of sons. The latter, when young or newly born, were commonly described by their fathers as "my hope" or "my consolation." Sons were seen as continuing a man's accomplishments, indeed his very character, into the future. Thus would a successful son reflect credit on his father—the credit of a "good name" or "good repute." However, these same effects might well reverse directions, with "the sins of the fathers" being found in the sons as well. Drunkards and gluttons, fornicators and thieves, would naturally reproduce themselves (*father as progenitor*).[8]

With so much at stake on both sides, it is not surprising that father-son relationships rested heavily on notions of "duty." Paternal oversight of education (in all aspects) was invariably portrayed in this light; sons, for their part, owed "respect" and "honor" in return. (A more tangible filial "duty" was assistance to fathers in their old age.) Yet this need not exclude "affection" of a deep and lasting kind. Affection and duty, affection energizing duty, duty controlling affection; such were the common formulas. Again, personal correspondence, when carefully read, displays the one no less than the other.

We should like, finally, to envision actual fathers and children caught up in the routine experience of everyday. Unfortunately, the records afford only random glimpses here and there; but these nonetheless are revealing. A father and his 10-year-old son carting grain to the mill; a father counseling his adult daughter on her impending marriage; a father and his son "discoursing" on witchcraft; a son and daughter joining their father in an argument with neighbors: from such small nuggets a cumulative picture emerges.[9] It is a picture at once consistent with the prescriptive materials cited previously and with our larger understandings of early American life. It is a picture, above all, of active, encompassing fatherhood, woven into the whole fabric of domestic and productive life. Indeed the critical point is that domestic and productive life overlapped so substantially—and were, in some respects, identical. The vast majority of seventeenth- and eighteenth-century fathers were, of course, farmers; of the remainder, all but a handful were local artisans and tradesmen. In either case, productive endeavor was centered upon the family hearth, and it seemed natural, even necessary, that children should be directly involved. From an early age boys and girls began to assist their father in the work of farm or shop. Fathers were thus a visible presence, year after year, day after day (*father as companion*). The same pattern obtained for leisure-time experience as well. Families attended church, went visiting, and passed the long hours of stormy winter days or summer evenings, largely as a unit. Fathering was thus an extension, if not an integral part, of much routine activity.

Additional aspects of "actual" fatherhood were rooted in demographic and ecological circumstance. In the first place, all adult men expected to become fathers, and only biological infertility might disappoint those expectations. (Perhaps 10 percent of colonial American marriages proved "barren.") Indeed, most would be fathers many times over. (An average couple produced about eight children surviving past infancy—though with some variation between one region or time period and the next). This, in turn, meant that fathering might continue up to, or into, old age. (Often a man was past 60 when his youngest child married and left home; of course, a considerable number did not live long enough to see that day.) Most fathers would have to suffer the loss of one or more of their children to some form of mortal illness.[10]

These statistical realities can be demonstrated with precision, but their meaning in emotional terms is far from clear. Some scholars contend that parents must have hedged, or limited, their investment in children whose prospects of survival were so uncertain.[11] Yet fragmentary evidence suggests otherwise. An occasional diarist or correspondent can be glimpsed in postures of extreme parental concern: for example, a prominent New England merchant who sat up and "watched" overnight whenever one of his children became seriously ill (*father as caregiver*). The depositional records of local courts afford scattered impressions of the same phenomenon: thus a village craftsman remembered that "when his child was sick, and like to die, he ran barefoot and barelegged, and with tears" through the night to find assistance.[12] Such materials raise, but do not resolve, important questions about the interior dimension of colonial fatherhood. Controversy will no doubt continue, and will hopefully spur additional research. In the meanwhile we should be wary of inference, from demography alone, that our forebears were uncaring toward their young.

II

The foregoing sketch of early American fatherhood is highly compressed and simplified. It describes at best, a set of mainstream trends or norms, without allowing for variance across space and time. That such variance was real and in some respects substantial cannot be doubted. "Puritan" fathers of seventeenth-century New England were not indistinguishable from their counterparts in the southern colonies or along the Apalachian frontier; nor were they exactly reproduced in their "Yankee" descendants of the following century. Even within the same region and the same time-period there might be important differences: one recent study posits coexisting but contrasting styles of "evangelical," "moderate," and "genteel" childrearing in colonial America.[13]

Yet no summary could possibly comprehend all such distinctions, and for present purposes the mainstream elements are the important ones. Almost everywhere fatherhood displayed the same active, integrated orientation. And in this there was little apparent change through the several generations of our "colonial period."[14]

When the story is carried forward into the national period, however, change appears as a central motif. It did not come all at once—indeed, its pace can easily be exaggerated—but its cumulative force seems, in retrospect, unmistakable. Indeed, the sense of change, with all its possibilities and perils, was widely manifest in the period itself.

The early decades of the nineteenth century brought a new burst of writing on domestic life. "Advice books" on courtship, on marriage, on homemaking, and above all on childrearing, fairly gushed from printing presses all across the land. Of course, advice of this type can never be directly equated with behavior, but it does—and did then—express prevalent attitudes and concerns. The concern of nineteenth-century writers on childrearing was partly a matter of nationalism: received (i.e., English) models were considered, in principle, unsuited to "republican" families. But something more was involved as well. The shrill tone of the new advice betrayed deep anxieties about the evolving shape and future prospects of the family. Change was actively embraced, and covertly feared, at one and the same time. And even as Americans envisioned new family forms, they wished to shore up the ways and values of the past.[15]

Significantly, this divided message was directed with special force toward mothers. *The Mother at Home, The Mother's Book, The Young Mother's Companion*: thus the titles of leading examples of the genre. Many of them contained a passage, or a chapter, on fathering; and there were companion volumes, at least a few, aimed primarily at fathers. However, the overall emphasis was clear—and was markedly different from the pattern of earlier times. Mother was now the primary parent. On her fell the chief responsibilities, more urgent and important than ever, for proper "rearing" of the young.[16]

These changes were underscored—indeed were prompted in part—by new ideas about gender. Virtually all human relationships were now reshaped by a massive system of (what modern sociologists would call) sex-role stereotyping. Women and men were thought to occupy different "spheres" appropriate to their entirely different *characters*. The female sphere was, of course, the home—and nothing more. Feminine character was calm, unselfish, in all ways "pure." It was woman's purity that, from a moral standpoint, elevated her far above men. It was also her purity that especially qualified her for motherhood.[17]

Ideas about human development were changing too, with convergent results for childrearing. Ministers, physicians, and advice writers increasingly stressed the formative influence of early—very early—experience. The Reverend Horace Bushnell, whose enormously popular treatise *Christian Nurture* was published in 1843, expressed the reigning wisdom on this point: "Let every Christian father and mother understand, when the child is three years old, that they have done more than half of what they will ever do for his character."[18] In fact, if not in so many words, such statements declared the transcendent importance of mothers. For mothers were now, more than ever, the leading caregivers to infants. The roots of maternal influence were seen as extending even *in utero*; thus numerous authorities believed that a mother's experiences during pregnancy might shape the destiny of her unborn child. (Did she wish, for example, to bear a future architect? Then she might spend her leisure hours gazing at fine buildings or pictures of buildings. Such, at least, was the strategy followed in one particularly famous instance—by the mother of Frank Lloyd Wright.)[19]

As mother's importance waxed, father's inexorably waned. Many of the leading advice-writers addressed this trend directly, disapproved it, and wished at the very least to retard its progress. But their words have the ring of special pleading, which only underscores the change. "I cannot believe," declared one clergy-man-author, "that God has established the relation of father without giving the father something to do."[20] Another urged simply that fathers "be careful not to under-rate [their] own duties or influence."[21] (Recall the quite similar words of caution—but about *mother's* role and influence—that had issued from comparable sources in the colonial period.)

And how were father's remaining "duties" described? First, and possibly foremost, he was still expected to set an official standard of morality for his family as a whole. He would conduct the family prayers (where that custom survived); he would lead in "edifying discourse" around the dinner table or fireside.[22] He was also the final arbiter of family discipline. Of course, in many routine matters mother's authority was sufficient, but when the stakes were high father must step in. These themes were frequently elaborated in domestic fiction—for example, a mother who threatens her recalcitrant child with the prospect of father's punishment ("when he returns home"). On other, happier occasions father would offer himself for play. He and his children might then enjoy "innocent games" or "romps" on the parlor floor.[23] Or he might simply observe and applaud; thus one father commented, in an actual case, "Three such rosy-cheeked children are not to be found as ours. . . . I wish you could have seen them this evening dance, while their mother played on the piano."[24]

Although fatherhood on these terms was hardly insubstantial, it diverged in obvious and important ways from the earlier pattern.

For one thing, it became part-time (*father as discussion leader, father as playmate*): for another, it opened some distance from the everyday workings of the household (*father as disciplinarian, father as audience*). The links between parents were reshaped accordingly. Thus a leading advice-book urged that fathers question their children every evening along lines such as the following: "Have you obeyed your mother in all things today? . . . Can you remember any case . . . in which you have tried to help your mother without her asking you to do so?"[25] And a popular work of domestic fiction described a woman's discussion with her husband about the "rearing" of their temperamental young son: "'Persevere, Anna, persevere,' were usually her husband's encouraging words. 'You are doing well. If anyone can mold right the disposition of the wayward child, it is you. I only wish that I had your patience and forebearance.'"[26] What actual women might make of such "encouragement" is difficult to say. But here lies another strand in the newly emergent pattern: *father as moral support*.

If distance and part-time involvement had come to characterize fatherhood (at least in some cases), it was preeminently for one reason. Beginning in the first decades of the nineteenth century, and increasingly thereafter, men were drawn out of their families toward income-producing work. The growth of large-scale commerce, of industry, and (later) of service enterprise—indeed, all the dynamic tendencies of economic modernization—yielded this overall effect: home and the workplace would no longer be the same. On the contrary, they became very different places, each with its own designs and purposes, its distinctive values and modes of activity. Around home was drawn an implicit but keenly felt boundary. On the far side lay "the world," including shops, factories, offices of all kinds. Passage across this boundary seemed difficult or dangerous, and must therefore be limited to necessary "occasions." Most of the latter involved men in their pursuit of "gainful occupation."[27]

The wrenching apart of work and home-life is one of the great themes in social history. And for fathers, in particular, the consequences can hardly be overestimated. Certain key elements of premodern fatherhood dwindled and disappeared (e.g., *father as pedagogue, father as moral overseer, father as companion*), while others were profoundly transformed (*father as psychologist, father as example*). Meanwhile new postures and responsibilities emerged from a general reordering of domestic life. Some of these are noted in the immediately preceding pages; but one remains as yet unmentioned—and is perhaps the most important of all. The "man of the family" going off to work and returning with the money needed to support an entire household, the "breadwinner," the resourceful fellow who "brings home the bacon": behind these now-tiresome cliches stands the figure of *father as provider*.[28] Of course, fathers had always been involved in the provision of goods and services to their families; but before the nineteenth century such activity was embedded in a larger matrix of domestic sharing. With modernization, it became "differentiated" as the chief, if not the exclusive, province of adult men. Now, for the first time, the central activity of fatherhood was sited outside one's immediate household. Now, being fully a father meant being separated from one's children for a considerable part of every working day.

That there was paradox, even painful contradiction, in this evolving pattern of experience seems plain in retrospect. But several generations were needed to work out its effects. At first, the "provider" role appeared to have enhancing implications for fatherhood. Providing could be seen and felt, on both sides, as an enlargement of paternal nurturance. The father who "brought home" the bacon, no less than the mother who cooked it and put it on the table, was supplying the vital needs of his children. What he actually did in his shop or office was little known to other members of his family; yet its product (income) *was* known, and was critical to their personal well-being. His ultimate "success" would depend on his strength of mind and will, his endurance, his moral fortitude—and perhaps on other qualities that women and children could but dimly imagine. As a result his work seemed mysterious and wonderful, and his ability to negotiate the treacherous routes through "the world" might be positively heroic.

In short, father's intrinsic connection to all that lay outside home gave him a special status within it. Whenever he returned to the "sacred hearth," he commanded attention, affection, respect, deference, devoted care. The sacrifices he had made, the risks he had run, the experience he had accumulated, the recognition he had achieved: all this made his opinions especially worthy to be heard (and accepted), his orders to be followed. A poem from midcentury may serve to exemplify such larger-than life fatherhood:

Father Is Coming

The clock is on the stroke of six,
    The father's work is done.
Sweep up the hearth and tend the fire,
    And put the kettle on:

The wild night-wind is blowing cold,
   'Tis dreary crossing o'er the world.

He's crossing o'er the world apace,
   He's stronger than the storm;
He does not feel the cold, not he,
   His heart, it is so warm:
For father's heart is stout and true
As ever human bosom knew . . .

Nay, do not close the shutters, child;
   For along the lane
The little window looks, and he
   Can see it shining plain.
I've heard him say he loves to mark
The cheerful firelight through the dark.

Hark! hark! I hear his footsteps now;
   He's through the garden gate.
Run, little Bess, and ope the door,
   And do not let him wait.
Shout, baby, shout! and clap thy hands,
For father on the threshold stands.[29]

Alas, such deeply idealized images are often problematic in relation to practice, and this one was especially so. It imposed, first of all, a standard of performance that only a portion of fathers could expect to meet. Success as provider did not come easily in nineteenth-century America. Popular formulas stressed "self-making," as noted previously; but in fact, opportunity was limited, if not foreclosed, by objective circumstances of race, class, ethnicity, place, and blind luck. And failure exacted a heavy price in self-, and social, esteem. To make matters worse, that price was shared with the families involved. A man who could not find his way in the world was likely to seem a failed father—in his own eyes, and in the eyes of those who mattered most to him.

How such disappointment was managed in individual cases is hard to imagine, and harder to study; but the rapid growth of the tramp population—a veritable "army," in the period phrase—suggests one (presumably extreme) line of response. Meanwhile, domestic fiction was full of failed fathers, their shortcomings as providers admixed with intemperance, improvidence, and plain malice. A common plot-line, for example, featured a hard-pressed father forcing his children to marry not for love but for money, as a way of relieving his own debts. The cost of these maneuvers was measured in broken hearts and otherwise blighted young lives, and death itself sometimes appeared as the ultimate result. One could well say of this literature that its most vivid portraits of fatherhood were destructive ones (*father as murderer*).[30]

Even for the best of providers there would be problems and pitfalls in fathering. Achievement in the world and constructive involvement at home: these made *separate* goals, often in conflict with one another. A key point of conflict was time: How much time could, or should, a busy father spend with his family? And, assuming that the quantity of such family time would be limited, what of its quality, its use in specific activities? Again and again prescriptive writings picked up the tension. Here is one fictive father, so driven by a "rapid increase in trade, and the necessity of devoting every possible moment to my customers" that he can no longer manage to conduct family prayers. (Subsequent events bring him back to his senses and his duty: "Better to lose a few shillings," he concludes, "than to become the deliberate murderer of my family, and the instrument of ruin to my soul.")[31] There is another, who cannot leave his business worries in the office and thus spoils the evening hour supposedly reserved for his children. (Dinner finds him "full of restless impatience, . . . hurrying through . . . in silence, . . . with an occasional suggestion to others to make the dispatch of which he sets so striking an example.")[32] At least occasionally, personal documents display the same conflict against the backdrop of actual experience. Thus a correspondent observed of Boston business families that "the points of contact between husbands and wives . . . were so few that a husband might 'become the father of a large family and even die without finding out his mistake.'"[33] And the wife of one such man complained (in a letter to relatives) that "his own business, and then that *dumb* committee, take every moment . . . Every eve'g he is out either at caucus or drinking wine with the Gov. at some gentry folks. So you see the evils consequent upon being a distinguished man."[34]

Of course, some fathers would manage to work through, or around, these demands on their time and energy. The advice literature furnished many hopeful models of "household management," designed to forge openings to fatherly interaction for even the busiest of men. Yet beyond the reach of any such contrivance lay deeper, more disabling problems. For sustained contact with "the world" touched men's innermost experience, indeed their very *character*. Many of the qualities most readily associated with success in work—ambition, cleverness, aggressive pursuit of the main chance—had no place in domestic life. At best, a man would have to perform an elaborate switch of role and behavior on crossing the threshold of his home. (It is notable that thresholds became a recurrent preoccupation in all sorts of writing from this period.) At worst, he would have to choose between effectiveness in one "sphere" or the other. And, given the convergent effects of practical need (i.e., for income) and cultural stereotype (i.e., about gender), most men instinctively preferred to concentrate on work and public affairs.

There was, moreover, an explicitly pejorative meaning to the contrast between home and the world. Home would exemplify the "purity" of women and the "innocence" of children. The world seemed deeply suspect from a moral standpoint—disordered, unstable, full of "traps" and "temptations" to vice in many forms. Personal integrity was largely discounted there. The struggle for success presented many inducements to abandon "true principle"; and one compromise led easily to the next. The men who lived and worked in this environment were necessarily imperiled, and maleness itself seemed to carry a certain odor of contamination.[35] This, too, greatly complicated father's role and whatever substantive contributions he might hope to make at home.

In order to strengthen his contributions and lessen the complications, father must fight down some of his deepest masculine tendencies. In doing so, he might well be guided by the example, if not the overt tutelage, of women. Thus men seemed characteristically "impatient of results," which was fine for business, but inappropriate to childrearing. Women, by contrast, were exemplars of patience, owing both to their "natural endowments" and to their cultural role and training. Similarly, men were given to sharp and severe methods of command, whereas women were better at the "gentle arts" of persuasion. Again, of course, it was women's way that best suited the home environment.

The prescriptive and fictional literature overflowed with illustrative cases. A typical story describes a troubled relationship of father and son. The father seeks an attitude of cheerful obedience to his generally reasonable commands. The son obeys, but not cheerfully. The father grows increasingly irritated, and the situation worsens—until a woman visitor, a certain "Aunt Mary," offers some helpful advice. Father's commands are usually given in "a cold, indifferent, or authoritative manner," but a "softer" approach, expressing "the sunshine of affection," might work to better effect. The father listens, and changes his tone; and his son's attitude improves remarkably.[36] Other stories have less happy endings. In one the father imposes on his timid and sickly son a harsh regimen of physical toughening—mostly outdoors, in wintertime, and late at night. Eventually the child dies, as much from "terror of his natural protector" as from overexposure and overexertion.[37] In a second story the father comes home "wearied and vexed" after "a hard day's labor," and finds his son covered with dirt and grime. Reacting too fast, he "taxes . . . and scolds [the boy] severely, . . ." only to learn a little later that the dirt has come from the performance of a particularly good deed. In the meantime the boy contracts pneumonia and dies before the father can make amends.[38] Such lugubrious tales expressed for nineteenth-century readers a profound and pervasive anxiety about fathering. Bad fathering was a matter not just of indifference or irrelevance, but of potentially deadly peril.

But this, we should remind ourselves, was the dark side of a highly variegated picture. Anxiety about fathering, fantasies of bad fathers: such things stand at some distance from the actual experience of individual families from one day to the next. Again we need behavioral evidence to set alongside the prescriptive and fictional materials; and again, such evidence is less available than we might wish. Still, current scholarship is beginning to sift and sort through quite voluminous quantities of personal documents from the period; and some early results can be summarized here in a tentative way.

It is apparent, first of all, that many fathers throughout the nineteenth century continued to have affectionate relationships with their children. Indeed their affection may now have been more openly expressed than in premodern times (when "moderation" of feeling was a touchstone of human relationship, and especially of childrearing). There is also much evidence of fatherly interest and involvement in the lives of children—insofar as other commitments allowed. Fathers continued to articulate for their sons and daughters an official code of conduct. They seemed to accept an implicit division of responsibility for the moral training of their young: the outside aspects fell chiefly to them (e.g., formulating general precepts, punishing major infractions), while mothers took care of the inner-life dimension (e.g., nurturing "conscience" and the growth of "steady habits"). Fathers also played an added role, with sons, in providing advice about "occupations" and public affairs (which, after all, were defined as exclusively male concerns).[39]

Yet, when viewed *in toto*, the behavioral evidence appears to confirm the trend toward limited fatherhood. Certainly it shows a shift in the relative weight of fathers' and mothers' parental contributions. Consider the following points of contrast with the pattern of premodern times:

1. *Then* fathers were the chief correspondents of their (adolescent and adult) children; *now* (i.e., during and after the mid-nineteenth century) mothers played that part at least as often.[40]
2. *Then* fathers played the central role in guiding, or fully controlling, the marital choices of their children. *Now* autonomy was the reigning principle in most aspects of courtship; and, to the extent that either parent was involved at all, it was usually the mother (and usually vis-à-vis her daughters).[41]
3. *Then* mothers seem to have been little concerned with any aspect of their sons' lives after childhood; *now* letters and diaries show them emotionally entangled with sons who were well into adulthood.[42]
4. *Then* it was common to give the largest share of blame or credit for adult outcomes to fathers; *now* the same judgments were made about mothers. "All that I am I owe to my angel mother": thus a favorite period cliche, echoed in one form or another by countless diarists and correspondents reflecting on their own childhoods.[43] Moreover, maternal influence was clearly associated with the special closeness of mothers and children. After all, wrote one man near the end of the century, "most people are on more confidential terms with their mother than with their father."[44]

The law of child custody presents one final way of tracing the same shift in parental roles and influence. Part prescription, part behavior, this evidence seems particularly vivid and compelling. In the premodern era, as noted above, custody belonged exclusively to fathers, reflecting their acknowledged status as the primary parent. However, this pattern was progressively altered by the courts of nineteenth-century America. At first, the changes were *ad hoc* and limited to cases of manifestly "unfit" (that is, "immoral" or "profligate") character. But in time the issue was more fully joined—and broadened. A Pennsylvania decision of 1810 rejected a father's claim to custody, for the reason (among others) that children of "tender age" need the "kind of assistance which can be afforded by none so well as a mother." A case in Maine two decades later elicited judicial opinions that: (1) daughters should be treated as "requiring peculiarly the superintendence of a mother," (2) sons "may probably be as well governed by her as by the father," and (3) "parental feelings of the mother toward her children are naturally as strong, and generally stronger, than those of the father." By 1847 a New York court could declare simply that "all other things being equal, the mother is the most proper parent to be entrusted with the custody of a child." To be sure, the older principles died hard—some courts asserted as late as 1834 that "in general, the father is by law clearly entitled to the custody of the child"—but the trend was more and more in the opposite direction. By the end of the century the law, no less than common opinion, affirmed maternal preeminence in childrearing.[45]

To speak of common opinion for any part of the nineteenth century is admittedly to oversimplify a complex and ever-changing situation. In fact, the United States of this era embraced a growing variety of groups and traditions, with a corresponding variance in family life. For those who lived in farm households, still a very large number, there was some carryover of earlier patterns. The image of the "family farm," validated in social and political terms by Jeffersonian ideology, lost little of its appeal; and where reality conformed to image, fatherhood might well retain an active, integrated orientation.

Immigrants made another important source of cultural and familial variance. Generalization about such a diverse mass is perilous; assuredly, all of the major ethnic groupings maintained their own styles and forms of domestic life. Still, it does seem clear that most nineteenth-century immigrants were of "peasant" background, and as such reflected a broadly premodern tradition. Integrated fatherhood was their way, too—or at least their expectation. However, the process of immigration was itself profoundly unsettling for family relations. Fatherhood, in particular, was tested and challenged by all the transitional steps from the "old country" to the new. Typically, fathers were responsible for decisions to migrate in the first place. Then, too, it was their responsibility to justify such decisions by converting American "opportunity" into tangible "success." Yet they were hampered again and again by their unfamiliarity with the language, the customs, the whole cultural ecology of the host country. Moreover, as fully formed adults they were necessarily limited in their adaptive capacities: younger people found it easier to learn and to change. More than their own children, immigrant fathers (and mothers) would remain "strangers in the land." They might even become a kind of embarrassment—the visible reminder of an outlived and outmoded past (*father as anachronism*). And this was an especially potent source of pain and tension, within families and across generations.[46]

Black families must also be mentioned here, if only to clear the record of inherited misunderstanding and prejudice. According to some accounts, black families were severely damaged, if not destroyed outright, first by slavery and later by the combined effects of rural poverty, mass migration, and the disordered environment of inner-city ghettos. The brunt is thought to have fallen with special force on black men—that is, fathers—who responded by simply opting out of domestic life: thus the myth of black "father absence." Black women were supposedly stronger, more authoritative, more responsive and responsible to family: hence the complementary myth of "black matriarchy."[47] But myths these were, and are. Recent historical scholarship shows a different picture altogether: a people that even in the midst of slavery incorporated many domestic values of the dominant culture (while retaining at least some "Africanisms"); a household system in which the presence, and leadership, of fathers was very much the norm. To be sure, some portion of black men cracked, or compromised, under the strain. But the vast majority were no less fully *fathers* than their white counterparts. Indeed, some may well have been more so. In their case, fatherhood carried special responsibilities for protecting the young (*father as shield*). Black children must first be guarded from, and then be armored for, the bitter shafts of racism.[48]

So it was that immigrant experience and black experience conditioned fatherhood in special ways. Yet, when seen in the longest perspective, these effects made for quantitative—not qualitative—difference from mainstream norms. Many families of certifiably "native stock" paid the price, in intergenerational conflict, for rapid social change. Each new cohort of children rendered its own fathers to some extent anachronistic. (Hence the implicitly pejorative term of reference for fathers: "my old man.") Similarly, most nineteenth-century fathers felt called on to protect their young from an uncaring world. True, the dangers were less immediate for white children than for black ones, but a fatherly shield would be needed all the same.

It should be clear, in any final view of the nineteenth century, that changing styles of fatherhood belonged most especially to a "modern" vanguard. The members of this group were largely white, Anglo-Saxon, and Protestant. They lived in cities. They worked in the new commercial and industrial "occupations." They belonged to what would later be called the "middle class." To them were directed the advice books and novels, from them came the diaries and letters, which have bulked so large in the present discussion. Relatively few at the outset, their numbers swelled prodigiously as the century passed, and they played in any case a style-setting role. Other groups observed, envied, resisted, adapted—and ultimately followed.

III

Many of the themes and tendencies just described were joined in a single cultural figure that has considerable resonance even today: the famous (or infamous) "Victorian patriarch." Here one feels the aura of authority surrounding nineteenth-century fatherhood, as well as the distance—even the danger—attaching thereto. The "patriarch" image also captures a certain isolation: father's elevated position exposed him to scrutiny by many others, not least by his own kith and kin. When, in premodern times, his responsibilities to family had been more immediately nurturant, he had been harder to see whole. Subsequently, however, he came out in the open—"available at last," as one study puts it, "for conscious examination."[49]

The process of examination, spanning the interval from the Victorian era to our own, has yielded much doubt and worry, and a steadily growing fund of direct criticism. The "patriarch" has been scaled down, and certain of his leading qualities have been modified. Still, in large measure he survives. Or at least the figure survives, and with it many consequences for actual experience. There is space in what follows only to outline this most recent part of the story.

In the first place, the end of the nineteenth century and the start of the twentieth produced a substantial elaboration of Victorian beliefs about gender. Maleness was defined more and more in terms of ambition and achievement, of what contemporaries called "push and go" or even "animal energy." As before, energy seemed invaluable in "the world" but irrelevant to family life, and the qualifier "animal" (sometimes rendered "animalistic") carried a menacing undertone.[50] Men must somehow be "tamed," or else be excluded from home and hearth. In his natural state the "male animal" could only disrupt the "family circle" (*father as intruder*).

The force of these images was augmented by change in the social environment. Work experience and domestic experience became ever more distinct, for greater and greater numbers of men. Large-scale, highly differentiated, "impersonal" organization was the pattern of choice in many forms of productive enterprise. Moreover, the late nineteenth century brought the first great boom in suburban living, as street railways (later buses and private automobiles) opened new vistas to "commuting."[51] Suburbs would soon become the epitome, in spatial terms, of the work-home dichotomy. "The suburban husband and father," noted one writer as early as 1900, "is almost entirely a Sunday institution."[52] Since such fathers spent so little time at home, they could not acquire savvy and skills in "domestic employments." They burbled and bumbled, and occasionally made fools of themselves. They were cajoled, humored, and implicitly patronized by long-suffering wives and clever children. Dagwood Bumstead (of the "Blondie" comic strip), Ozzie Nelson (of the popular radio show "Ozzie and Harriet"), and the faintly ridiculous hero of Clarence Day's memoir (later a Broadway play) *Life with Father* made well-known variants on the general type (*father as incompetent*).

The snickering attitude that infused this imagery betrayed as well a deeper tension in father-child relations. With family size greatly shrunken (as compared to pre-modern times), with family boundaries ever more tightly drawn, with family values expressed more and more in psychological terms, domestic relations assumed a particularly intensive form. It was no accident, for example, that Freudian theory, with its cornerstone concept of the "Oedipus complex," found its most receptive hearing in the United States.[53] To be sure, intensive families were increasingly common throughout the Western World by the early twentieth century; but American experience created special, and amplifying, effects. One of these was "anachronistic" fatherhood, as noted above: fathers who seemed "old-fashioned" alongside their more open and adaptable children. In fact, such differences acquired a normative significance from progressive views of historical development. The hope of America, so widely and fervently proclaimed, was its future—which was to say its "younger generation." Other aspects of cultural ideology also played in here—most especially the "cult of success" and the manifold encouragements to social mobility.[54] Success was measured in terms of the distance from the starting point to the finish in any given life, and the starting point was the position of father. Here lay an inducement to competition between the generations, powerful and pervasive—however covert—and uniquely American (*father as rival*).

Does this seem too harsh, this portrait of intrusive, incompetent, and competitive fatherhood? In fact, there were other elements that must also be weighed in the balance. Most important perhaps was the continued salience of "breadwinning" in men's personal and domestic experience. The image of *father as provider* was, if anything, stronger in the opening decades of the new century than ever before. Moreover, it conferred on many individual fathers a special status, expressed in attitudes of respect, of deference, of grateful love on the part of other family members.

In addition to exploiting these traditional advantages, twentieth-century fathers have also accepted—have created, in part—one new role of a "softer" sort. In the hours left over after their work and public duties—on evenings, weekends, and vacations—they have sought to engage their children in a variety of comradely activities. Much of this is by way of recreation (attendance at movies or sports events, camping, fishing); some involves projects of household maintenance or improvement (gardening, redecorating, repair work). One hears echoes of the old "companionate" theme in father-child relations, but with a difference. The modern pattern is more contrived and self-conscious, and altogether more confined. For example, it is confined chiefly, or only, to sons; the relation of fathers and daughters, by contrast, has no clear focus and little enough substantive content. Nonetheless, shared recreation can be—often is—deeply meaningful on both sides. The aim is to discount for a time differences of age, of experience, of status—to find some neutral ground on which father and son can meet more or less as equals (*father as chum*). Perhaps this underlies the spread of a relatively new term of address for fathers, the now almost ubiquitous "dad" or "daddy." There is a note of affectionate familiarity here. But there may also be some implicit tendency to patronize: "dad" slides easily into "poor dad."

This brief sketch of developments in our own century has scarcely touched "events" as such. Did the Depression, then, have no effect on fathers *qua* fathers? And the two world wars? And the political turmoil of still recent memory (with its specifically "generational" confrontations)? A short answer would have to be that such events did most certainly affect many individual men in their personal experience as fathers, but did not alter fatherhood as a category of social experience. The Depression attacked, and sometimes shattered, fathers in their central role as providers; but the role itself survived until the return of better times, and flourished thereafter. The wars separated millions of fathers from their families for months or years at a stretch, but the ensuing peacetimes brought a renewal (even a reinforcement) of traditional domestic arrangements. The reform and countercultural movements of the 1960s challenged authority of many sorts—including, and especially, the paternal sort—but with little long-term effect.

Still, this is not to say that fatherhood is wholly uninfluenced by larger currents of change. And two changes, recently begun and still in progress, deserve special notice here. Both bear strongly on the experience of individual fathers; either, or both, may alter the category as well. The first is the entry of women—most strikingly, of married women with small children—into the working world outside the home. The second is the growing incidence of divorce and thus of single parenthood (even, in a small portion of cases, of single *father*hood). As a result of these trends the vast gulf between the experiences of men and women, anchored in more than a century of our history, has finally begun to close. Also, and paradoxically, divorce has led in some cases to an enlarged experience of fatherhood (even where the children involved live primarily with their mother). Indeed, maleness itself (including fatherhood) is increasingly subject to reconsideration of a very elemental sort. The old verities about hard, striving, emotionally invulnerable men seem dubious in fact and distressing in result. And it is precisely these forces that have limited the range and depth of fathering in the past.[55]

To be sure, the same trends have a dark underside. For whatever reason—the strain implicit in the new sex roles, the freedom of easy divorce, the confusion around the meanings of masculinity— some men are avoiding or abandoning fatherhood altogether. The considerable number who decline to have children in the first place, and the far greater number who marry (or not, as the case may be), procreate, leave, and refuse to supply even the most minimal levels of "child support" (*father as abdicator*): these facts, too, must be weighed in the present-day balance.

Can history help us to visualize the fathers of the future? Could it even yield "lessons" of some value in coping with that future? No lessons in the literal sense, but, for those who would welcome further change, a measure of hope—and a caution.

The hope is founded on the plain fact that received models of fatherhood are not writ in the stars or in our genes. Our ancestors knew a very different pattern from our own, and our descendants may well have another that is no less different. Fatherhood, history reminds us, is a cultural invention.

The caution is that all such inventions are deeply rooted in contemporaneous structures of society and culture, of belief and custom, and even of "depth psychology." Thus change in the role, broadly conceived, can only be slow, incremental, painful. Change in individuals may come more quickly, and go deeper—but still within some limits.

As always, history allows us to be hopeful and compels us to be humble, at one and the same time.

NOTES

1. See, for example, E. Anthony Rotundo, "American Fatherhood: A Historical Perspective," in *American Behavioral Scientist, XXIX* (1985), 7–25.

2. Cotton Mather, *The Diary of Cotton Mather*, ed. Worthington Chauncey Ford, 2 vols. (repub. New York, 1969), I, 239–40.

3. See Ruth H. Bloch, "American Feminine Ideals in Transition: The Rise of the Moral Mother, 1785–1815," *Feminist Studies*, IV (1978). Explicit statements of this viewpoint can be found, for example, in Puritan sermon literature. Thus Rev. Samuel Willard of Boston believed that while both parents have "a share in the government of [their children] . . . there is an inequality in the degree of this authority, and the husband is to be acknowledged to hold a superiority, which the wife is practically to allow." (Quoted in Edmund Morgan, *The Puritan Family: Religion and Domestic Relations in Seventeenth-Century New England*, rev. ed. [New York, 1966], 45.) And an English Puritan, whose writings were read on both sides of the Atlantic, declared on the same subject: "A wife may not simply without, or directly against her husband's consent, order and dispose of the children in giving them names, apparelling their bodies, appointing their callings, places of bringing up, marriages, or [inheritance] portions." (William Gouge, *Of Domesticall Duties* [London, 1622], 309.) For an elaborate, and important, study of the "controlling" aspects of early American fatherhood, see Philip J. Greven, *Four Generations: Population, Land, and Family in Colonial Andover, Massachusetts* (Ithaca, N.Y., 1970).

4. John Demos, *A Little Commonwealth: Family Life in Plymouth Colony* (New York, 1970), 82–84; Morgan, *The Puritan Family*, 84ff; Laurel Thatcher Ulrich, *Good Wives: Image and Reality in the Lives of Women in Northern New England, 1650–1750* (New York, 1982), 6ff, 106–10; Lyle Koehler, *A Search for Power: The "Weaker Sex" in Seventeenth-Century New England* (Chicago, 1980),

28–61; John Winthrop, *Winthrop's Journal, 1630–1649*, ed. James Kendall Hosmer, 2 vols. (New York, 1908), II, 225; Demos, *Entertaining Satan: Witchcraft and the Culture of Colonial New England* (New York, 1982), 60–64; Koehler, *A Search for Power*, 276–81.

5. These points are effectively developed in Bloch, "American Feminine Ideals in Transition." See also Ulrich, *Good Wives*, 154–56.

6. See Mary Beth Norton, *Liberty's Daughters: The Revolutionary Experience of American Women, 1750–1800* (Boston, 1980), 94–100; and Daniel Blake Smith, *Inside the Great House: Planter Family Life in Eighteenth-Century Chesapeake Society* (Ithaca, N.Y., 1980), chs. 1, 3; Michael Grossberg, "Law and the Family in Nineteenth-Century America," unpub. Ph.D. diss. (Brandeis University, 1979), 257–59.

7. Samuel Gay, letter to Ebenezer Gay, 29 March 1809. (Quoted in E. Anthony Rotundo, "Manhood In America: The Northern Middle Class, 1770–1920," unpub. Ph.D. dissertation, Brandeis University, 1981.)

8. See Rotundo, "Manhood in America, 1770–1920."

9. The examples mentioned here, and many others like them, can be gleaned from the depositional files of early American courts. See, for instance, *Records and Files of the Quarterly Courts of Essex County, Massachusetts*, 8 vols. (Salem, Mass., 1911–21), *passim*.

10. The demographic research, on which these generalizations are based, can be sampled in Demos, *A Little Commonwealth*, 65–68, 192–93, and in Demos, "Old Age in Early New England" (chapter seven in the present volume).

11. See, for example, Edward Shorter, *The Making of the Modern Family* (New York, 1975), ch. 5, especially pp. 203–4; and Lawrence Stone, *The Family, Sex and Marriage in England, 1500–1800* (New York, 1977), 70, 105–6.

12. Samuel Sewall, *The Diary of Samuel Sewall*, ed. M. Halsey Thomas, 2 vols. (New York, 1973), I, 88–89, 266, and *passim*; Samuel G. Drake, *Annals of Witchcraft in New England* (New York, 1869), 239.

13. Philip J. Greven, *The Protestant Temperament: Patterns of Child-Rearing, Religious Experience, and the Self in Early America* (New York, 1977).

14. A possibly significant exception should be noticed here. Plantation-owners in the southern colonies during the eighteenth century seem—according to one account—to have maintained an attitude of "indifference to [their] children, or, at any rate, . . . uninvolvement in their rearing." See Michael Zuckerman, "Penmanship Exercises for Saucy Sons: Some Thoughts on the Colonial Southern Family," *South Carolina Historical Magazine*, LXI (1982), 152–66; also Zuckerman, "William Byrd's Family," *Perspectives in American History*, XII (1979), 253–311. This viewpoint is, however, disputed by other scholars; e.g. Smith, *Inside the Great House*, chs. 1, 3.

15. Robert Sunley, "Early Nineteenth-Century American Literature on Child Rearing," in Margaret Mead and Martha Wolfenstein, eds., *Childhood in Contemporary Cultures* (Chicago, 1955), 150–67; Bernard Wishy, *The Child and the Republic: The Dawn of Modern American Child Nurture* (Philadelphia, 1968).

16. See Bloch, "American Feminine Ideals in Transition," and Linda Kerber, *Women of the Republic: Intellect and Ideology in Revolutionary America* (Chapel Hill, 1980), ch. 7.

17. Barbara Welter, "The Cult of True Womanhood: 1820–1860," *American Quarterly*, XVIII (1966), 151–74; Nancy F. Cott, *The Bonds of Womanhood:*

Woman's Sphere in New England, 1780–1835 (New Haven: Yale University Press, 1977); Carl N. Degler, *At Odds: Women and the Family in America from the Revolution to the Present* (New York, 1980), 73ff.

18. Horace Bushnell, *Christian Nurture* (New York, 1843), 48.

19. Frank Lloyd Wright, *An Autobiography* (New York, 1932), 8.

20. William A. Alcott, "Woman But a Helper, Designed for Fathers," in *Parents' Magazine* (December, 1841), 88–89; quoted in Rachel Deborah Cramer, "Images of the American Father, 1790–1860," unpublished honors thesis (Brandeis University, 1980), 45.

21. Theodore Dwight, *The Father's Book* (Springfield, Mass., 1835), iii.

22. See Cramer, "Images of the American Father, 1790–1860," 51–52.

23. *Ibid.*, 50.

24. Letter from Sylvester G. Lusk, to Sylvester Lusk, 31 Jan. 1840; quoted in Rotundo, "Manhood in America, 1770–1920."

25. John S. C. Abbott, "Paternal Neglect," *Parents' Magazine* (March 1842), 149; quoted in Cramer, "Images of the American Father, 1790–1860," 52.

26. T. S. Arthur, *The Mother* (Boston, 1846), 25.

27. This theme is powerfully developed, with particular reference to the experience of nineteenth-century women, in Degler, *At Odds*. See also Cott, *The Bonds of Womanhood*, Welter, "The Cult of True Womanhood, 1820–1860," and Christopher Lasch, *Haven in a Heartless World* (New York, 1977). It is possible, however, to exaggerate the extent of the home/world dichotomy, and the speed with which it developed; see Elizabeth H. Pleck, "Two Worlds in One," *Journal of Social History*, X (1976), 178–95.

28. This role is underscored, for example, in marriage manuals from the period. See Michael Gordon, "The Ideal Husband as Depicted in the Nineteenth-Century Marriage Manual," in Elizabeth H. Pleck and Joseph Pleck, eds., *The American Man* (Englewood Cliffs, N.J., 1980), 145–57.

29. From Mary Howitt, *The Children's Hour* (January 1868), quoted in Cramer, "Images of the American Father, 1790–1860," i.

30. See Cramer, "Images of the American Father, 1790–1860," 79–84.

31. "Family Prayer by Men of Business," in *Parents' Magazine* (May 1842), 198ff.; quoted in Cramer, "Images of the American Father, 1790–1860," 35.

32. "The Father," in *Parents' Magazine* (April 1842), 174; quoted in Cramer, "Images of the American Father, 1790–1860," 48.

33. Quoted in James R. McGovern, *Yankee Family* (New Orleans, 1975), 85.

34. Letter from Sarah Russell, to Thomas Russell, 16 Jan. 1845; quoted in Rotundo, "Manhood in America, 1770–1920," 178.

35. On this point, see Carroll Smith-Rosenberg, "Sex as Symbol in Victorian Purity: An Ethnohistorical Analysis of Jacksonian American," in John Demos and Sarane Spence Boocock, eds., *Turning Points: Historical and Sociological Essays on the Family* (Chicago, 1978), 212–47.

36. T. S. Arthur, "Aunt Mary's Suggestion," in *Home Lights and Shadows* (New York, 1854), 249; summarized and quoted in Cramer, "Images of the American Father, 1790–1860," 33–34.

37. Lydia Sigourney, "The Intemperate," in *Sketches* (Amherst, Mass., 1834), 178ff.; summarized and quoted in Cramer, "Images of the American Father, 1790–1860," 79–80.

38. H. A Graves, "The Parent's Loss," in *The Family Circle* (Boston, 1845),

85; summarized and quoted in Cramer, "Images of the American Father, 1790–1860," 81.

39. Evidence on these points is presented in Rotundo, "Manhood in America, 1770–1920," *passim.*

40. Ibid., *passim.*

41. See Ellen K. Rothman, *Hands and Hearts: A History of American Courtship* (New York, 1984), 26–30, 116–18, 216–18.

42. Rotundo, "Manhood in America, 1770–1920," *passim.*

43. Bloch, "American Feminine Ideals in Transition," 116ff.; John Demos, "The American Family in Past Time," *The American Scholar*, XLIII (1974), 441ff.

44. Letter from Howard Taylor Ricketts, to Myra Tubbs, 11 July 1896; quoted in Ellen K. Rothman, "'Intimate Aquaintance': Courtship and the Transition to Marriage in America, 1700–1900." Ph.D. diss. (Brandeis University, 1981), p. 200.

45. Jamil S. Zinaldin, "The Emergence of a Modern American Family Law: Child Custody, Adoption, and the Courts, 1796–1851," in *Northwestern University Law Review*, LXXIII (1979), 1038–89.

46. The fullest, deepest, treatment of this theme is still Oscar Handlin's classic study, *The Uprooted* (Boston, 1950); see esp. ch. 10.

47. E. Franklin Frazier, *The Negro Family in the United States* (repr. Chicago, 1966); Daniel P. Moynihan, "The Negro Family: The Case for National Action," in Lee Rainwater and William L. Yancey, eds., *The Moynihan Report and the Politics of Controversy* (Cambridge, Mass., 1967), 41–124.

48. Herbert G. Gutman, *The Black Family in Slavery and Freedom 1750–1925* (New York, 1976); Eugene Genovese, *Roll, Jordan, Roll: The World The Slaves Made* (New York, 1974), 482–94.

49. Fred Weinstein and Gerald Platt, *The Wish to be Free: Society, Psyche, and Value Change* (Berkeley, 1969), 147.

50. Joe L. Dubbert, *A Man's Place: Masculine Identity in Transition* (Englewood Cliffs, N.J., 1979), ch. 5; Rotundo, "Manhood in America, 1770–1920," *passim*; James R. McGovern, "David Graham Phillips and the Virility Impulse of the Progressives," *New England Quarterly*, XXXIX (1966), 333–48.

51. Sam Bass Warner, Jr., *Streetcar Suburbs* (Cambridge, Mass., 1963). See also Dubbert, *A Man's Place*, 140ff.

52. "Rapid Transit and Home Life," in *Harper's Bazaar*, XXXXIII (Dec. 1900), 200; quoted in Peter Gabriel Filene, *Him/Her Self: Sex Roles in Modern America* (New York, 1975), 77.

53. John Demos, "Oedipus and America: Historical Perspectives on the Reception of Psychoanalysis in the United States," in *The Annual of Psychoanalysis*, VI (1978), 23–39.

54. John J. Cawelti, *Apostles of the Self-Made Man* (Chicago, 1965); I. G. Wyllie, *The Self-Made Man in America: The Myth of Rags to Riches* (New Brunswick, N.J., 1954).

55. The recent literature of the "men's liberation movement" is, of course, voluminous. A brief introduction, with references to other works, can be found in Pleck and Pleck, *The American Man*, 37–40, 417–33.

CHAPTER IV

Child Abuse in Context: An Historian's Perspective

In the late 1970s came another sort of inter-disciplinary invitation. The National Alliance for the Prevention and Treatment of Child Abuse had received foundation support for a broad reappraisal of its work and priorities "from the vantage-point of the humanities." I was one of a half-dozen "humanists" asked to take part. (The others included philosophers, literature specialists, and art historians.)

The means to this end were familiar enough: we would write "position papers," hold public conferences, and plan (sometime in the future) the publication of a book. But the end itself seemed distressingly vague. Indeed I felt an acute sense of discrepancy just here: between the very palpable, and painful, and contemporary facts of "the child abuse problem" and the facts in my head about Life Long Ago. I wondered, more than once: perhaps the foundation money would be better spent on bandages for battered babies or psychotherapy for their parents?

It was easier to avoid this question than to answer it. But finally, too, I did persuade myself that the history of child abuse furnished a certain encouragement about its present-day status. A social problem that has always "been there" may reasonably be regarded as intractable. But a problem that did not exist at one or another point in the past may again cease to exist in the future. On child abuse, I opt for the latter—more hopeful—viewpoint.

In addition, this project drew me into a large and lively schol-ars' debate. It is now fashionable, in family history, to believe that things have been getting better and better as the centuries move along, that the passage of time has increased tender feeling—and diminished indifference, or outright cruelty—between spouses, be-tween parents and children, between family-members of all sorts. (In one rendition the formula is "love-up, anger-down.") Among other things, the essay that follows is a way of saying "I doubt it."

Child abuse—as a public issue—has a very short history. A single event marks its beginning: the publication in 1962 of an article entitled "The Battered Child Syndrome" in the *Journal of the American Medical Association*.[1] The impact of this article owed something to authorship (five highly respected physicians headed by C. Henry Kempe of the University of Colorado Medical Center), something to provenance (one of the leading medical journals in the country), and not a little to rhetorical inventiveness (e.g. the term "battered child," which soon gained wide currency among professionals and lay persons alike).

There was, in addition, the intrinsic power of the issue itself. Child abuse evoked an immediate and complex mix of emotions: horror, shame, fascination, disgust. Dr. Kempe and his co-authors noted that physicians themselves experienced "great difficulty . . . in believing that parents could have attacked their children" and often attempted "to obliterate such suspicions from their minds, even in the face of obvious circumstantial evidence." In a sense the problem had long been consigned to a netherworld of things felt but not seen, known but not acknowledged. The "Battered Child" essay was like a shroud torn suddenly aside. Onlookers reacted with shock, but also perhaps with a kind of relief. The horror was in the open now, and it would not easily be shut up again.

From the pages of professional journals discussion of child abuse spread quickly to numerous public settings. There was strong response from the news media, from charitable organiza-tions, and from government at every level. Research monies poured forth to a variety of individual investigators; the resultant reports and recommendations soon amounted to a huge scientific literature. Elected officials in state after state rushed to create legislation that would strengthen protective services to children. Legal proceedings were undertaken, and judicial decisions ren-dered, with the aim of clarifying both public and private responsibilities in this area. The energies of the federal government were also enlisted in the cause: a Child Abuse Prevention and Treatment Act emerged (in 1973) as a direct result, as did a new National Center on Child Abuse and Neglect.[2]

Of course, the long-term outcome of these activities remains uncertain. Does society have the will and the means to eradicate child abuse, or at least to reduce its incidence? What substantive strategies show the greatest prospect of success? What indeed is the most appropriate framework for *understanding* the problem—and for taking effective remedial measures? Such questions are still well short of resolution.

In the meantime a second group of questions, different in character but not unrelated, reach out from the fact of child abuse to its cultural and historical context. Is there something inevitable and irreducible about all this, regardless of particular social settings? Must we assume a certain *residuum* of abusive impulse— and behavior—in any given human population? Does the incidence of such behavior vary significantly when measured across a range of human cultures? And, within the history of our own culture, do we find child abuse as a continuous presence or a variable one (depending, in short, on time and place)?

The present essay addresses the last of the above questions. This will be an historical inquiry, framed by explicit comparisons of our present with our past. Unfortunately, no systematic research on the history of child abuse is currently available in print (though this has not stopped some writers from pronouncing strong and sweeping opinions on the subject). Hence the ensuing pages unfold an argument that is necessarily tentative and incomplete. The conclusions proposed here would need careful testing, by many hands, in order to become solidly persuasive. Yet all historical inquiry is incremental, and every scholarly project must begin from "inadequate data." Indeed, as we struggle by various routes toward a reckoning with child abuse, history has strong claims on our attention. For history, however imperfectly understood, is one of our best aids in discovering what kind of a people we are.

I

While the aforementioned article by Dr. Kempe and his colleagues produced a first dramatic surge of public recognition, we should not assume that child abuse had no importance before 1962. Recognition and reality (i.e. *what* is recognized) rarely march in lock step; indeed it is the task of this essay to analyze points of disjunction between the two.

It seems clear in retrospect that the battered child was "discovered" by a series of sequential steps. Of great importance, for example, was the perfection of scientific technique in the medical sub-specialty of pediatric radiology. This enabled investigators such as John Caffey (of the College of Physicians and Surgeons at Columbia University) and F. N. Silverman (of the University of Cincinnati College of Medicine) to analyze with new evidence a puzzling "syndrome" of "infant trauma"; their findings were published in several technical articles beginning in the late 1940s. In fact, physicians had observed some elements of this pattern as long as a hundred years ago: swellings, bruises, and fractures, in various forms and combinations. They had not, however, managed to pinpoint its source; rickets was their usual diagnosis—"in the absence," noted one commentator, "of any other assignable cause."[3]

The earliest of these medical descriptions date from the closing decades of the last century. Meanwhile churches and charitable organizations had begun to move on a parallel track. A New York girl named Mary Ellen, discovered savagely beaten by her (adoptive) parents in the year 1874, is often cited as the "first recorded case" of child abuse. Mary Ellen was removed from her home following a lawsuit brought by the American Society for the Prevention of Cruelty to Animals. (The law of this era afforded more in the way of formal protection to animals than to children as such.) Succeeding years brought the establishment of various Societies for the Prevention of Cruelty to Children, whose self-defined concerns included child abuse (roughly as we would understand the matter today).[4]

But this is about as far as we can follow the historical trail of child abuse, on the basis of clearly identified markings. Before the middle of the nineteenth century the evidence is thin and vague and very scattered. Fictional writings, such as the novels of Charles Dickens, are sometimes cited in this connection, but their relation to social reality is inherently problematic. (And they would, in any case, push the story no more than a few decades further back.)

So what are we to conclude about child abuse in all the years before, let us say, 1800? "Recorded cases" are not to be found; shall we therefore assume an absence of actual *behavior*? To this question most investigators return a negative answer. The change, they assert, lies not with the behavior but with the records—and with the social responses which record-keeping reflects. In brief, their argument goes as follows. Children have always been abused (and perhaps more widely abused, the further one looks back in time). What history has added, over roughly the past century, is a new sensibility—a feeling of outrage that things should be so. The various behaviors which we now designate as child abuse were widely prevalent, were taken for granted, and were scarcely (if ever) discountenanced, by countless generations of our forebears.

This way of thinking is apparently common among the various professionals active in the child abuse "field" today. One such person, a physician and author of numerous books and articles, writes flatly: "The neglect and abuse of children has been evidenced since the beginning of time. The natural animalistic instincts of the human race have not changed with the passing of the centuries." Another declares that "maltreatment of children has been justified for many centuries by the belief that severe physical punishment was necessary either to maintain discipline, to transmit educational ideas, to please certain gods, or to expel spirits." Still another notes "a discrepancy between the magnitude . . . of child abuse in history and its documentation," but then proceeds, undeterred, to paint a particularly grisly picture of his subject.[5] In fact, academic historians have been no less inclined to these same assumptions. "Of course," writes one scholar, "a great deal of casual wife-beating and child-battering, which today would end up in the courts, simply went unrecorded in medieval and early modern times." And a second declares his firm belief that "a very large percentage of the children born prior to the eighteenth century were what would today be termed 'battered children'."[6]

In turning to the historians, we enter a very broad area of inquiry—where child abuse as such occupies only a small corner. Historical research has recently broached the subject of family life in a massive way, and the outlines of a "consensus view" are becoming increasingly clear. According to this view the history of family life in general, and of childhood in particular, has a markedly progressive cast. Change runs from "the Bad Old Days" (as one scholar has chosen to call them) toward something substantially better: from indifference and brutality and emotional constriction, toward kindliness and closeness and a burgeoning spirit of "affective individualism." Here is the old "whig interpretation of history" in a new guise; here, in short, is an (ultimately) happy story.[7]

But wherever it focusses on pre-modern childhood, the story is anything *but* happy; a recent and influential statement begins as follows. "The history of childhood is a nightmare from which we have only recently begun to awaken. The further back in history one goes, the lower the level of child care, and the more likely the children are to be killed, abandoned, beaten, terrorized, and sexually abused." Subsequent parts of the same essay unfold a theory of "parent-child relations" through history, based on a sequence of characteristic "modes." The gist can be grasped quite readily from the labels themselves: (1) Infanticidal Mode (Antiquity to Fourth Century A.D.); (2) Abandonment Mode (Fourth to Thirteenth Century A.D.); (3) Ambivalent Mode (Fourteenth to Seventeenth Centuries); (4) Intrusive Mode (Eighteenth Century); (5) Socialization Mode (Nineteenth to Mid-Twentieth Century); (6) Helping Mode (Begins Mid-Twentieth Century).[8]

There is insufficient space here to examine this construction at length (or, indeed, the constructions of other, like-minded investigators). However, it may be worthwhile to cite a few particular specimens of evidence and inference, if only to exemplify the larger corpus to which they belong.

Item: Parental reference to children in diaries, correspondence, and other personal documents from pre-modern times was sparse, brief, sometimes laconic. Evidently this reflected an attitude of profound indifference to the fate of one's own offspring.

Item: Two children, within a single family, were sometimes given the same name (usually after the older one had died). This implies a lack of individualized attachment to either one.

Item: Pre-modern child-rearing included, among its favored techniques of discipline, whipping, threats of death and damnation, invocation of ghosts, goblins, witches, and the like. There was, then, no appreciation of the tender sensibilities of the young.

Item: A reliance on "child labor" (most shockingly, during the early phases of the Industrial Revolution) was well nigh ubiquitous. Here we observe an attitude not just of indifference but also of rank exploitation.

Item: In at least some places, at some points in history, infanticide seems to have been practiced—even "accepted"—on a fairly broad scale. This evinces callousness of the most extreme sort.

Item: In certain pre-modern settings—those associated with "Puritanism," for example—infants were viewed as inherently corrupt and depraved; and parental duties were framed in terms such as "breaking the [child's] will." Hence the underlying *disposition* toward children was unfriendly, to say the least.

II

Such numerous and varied materials seem at first glance to offer powerful support for the "nightmare" view of the history of childhood. Yet a second (longer) look gives reason for pause. Personal documents, for example, are probably the wrong place to search for signs of affect toward children, for pre-modern diarists and correspondents did not often write about *any* sector of their affective experience. The practice of naming newborn children for deceased siblings may have reflected a deep sense of loss, and a corresponding wish for "replacement." Much of pre-modern child-rearing does indeed seem harsh by our standards, but it was not incongruous, and certainly not capricious, given then prevalent views of human nature and society. (Moreover, the chill undercurrents in traditional fairy tales may serve the most basic developmental needs of children; such at least is the argument of the psychologist Bruno Bettelheim in his fascinating study *The Uses of Enchantment.*[9]) Child labor was, before the Industrial Revolution, a relatively benign affair. Admittedly, the development of the factory system introduced new and dangerous elements, but the involvement there of children was, historically speaking, of limited duration. Infanticide appears to have been closely associated with illegitimacy; in some (most?) of the documented cases, it was the desperate recourse of unmarried mothers faced with both material privation and social stigma. Finally, the beliefs which emphasized the inborn wickedness of children also furnished hope for their redemption; in a sense parental repressiveness would redound to the eternal benefit of the "little sinners."

Taken altogether, these considerations underscore the importance of *context* in the evaluation of any given piece of social behavior. The overall demographic regime, the material basis of life, the prevailing system of beliefs and values, the intrinsic limitations of the evidence itself: all this, and more, must be carefully weighed in the interpretive balance. Of course, no less applies in the evaluation of variant practice in our own time; what seems "deviant" or downright "abusive" from one vantage point may appear innocent, or at least of good intent, from another. The situation of a child left unattended along a busy city street cannot be equated with that of his age-mates turned loose in a suburban patio or rural barnyard. The ostensible reasons for the parental behavior may be the same in each case—say, a trip to the supermarket—but the *meaning* depends on the total configuration of circumstance. Similarly, the use of physical force (e.g. spanking) in disciplining children yields to no single standard of judgment. Is such practice routine in the family (and community) of the children involved? What is the prevalent pattern of physical expressiveness generally? And (most obviously) how severe, or possibly injurious, are these disciplinary techniques in their actual application?

Clearly, the task of identifying child abuse must be approached with caution. And, in the remainder of the present essay, we must accept a rather minimal—that is, narrowly circumscribed—definition of terms. In this way we may hope to reduce the uncertainties relating to contextual influence. We seek, in short, a baseline definition of child abuse to which most, if not all, observers could give assent. Four elements seem essential: (1) *physical force*, applied (2) *intentionally*, so as (3) to inflict *substantial, even life-threatening, injury* on the body of the child, by (4) *his/her own parents.*[10] In asserting these criteria, we exclude other possibilities. The large and varied area of "neglect"—the failure to provide for the central needs of the child—is left aside, for neglect seems a particularly elastic and elusive category, permeable by all manner of cultural influence. Harsh discipline (which falls short of causing injury) and a seeming lack of empathy for childhood also fall outside our definitional boundaries. Finally, we exclude the abusive treatment of children by persons other than their own parents. This last may seem somewhat arbitrary, yet it does correspond with common usage. The particular horror of modern-day child abuse is its association with adults who might otherwise be seen as the chief protectors of the victims. And the particular shadow which it casts on the contemporary social order is the rending of these most necessary and "natural" of all human ties.

But in order to make a fresh start in the historical study of child abuse, we need not only a clear understanding of terms but a new research strategy as well. Prior comment on this subject has been muddled by a certain looseness of time perspective—a tendency to lump together disparate fragments from many epochs and cultures. Generalizations fashioned by such means are inherently suspect. Hence in the pages that follow we shall assay a radically different approach. We shall focus on a single historical setting, sacrificing breadth for depth in the development of a "case-study." This will enable us to control more effectively for context (as previously discussed) and for the inherent vagaries of the evidence itself. Our conclusions will express obvious limitations of time and place, but within those limitations should attain definitive shape. Moreover, they will not lack *implications* of a broader sort, and may eventually lend themselves to useful comparison with other historical "cases."

The particular focus to be developed here is the culture of early America—more specifically, of early New England—a culture distinctly "pre-modern," yet related to our own by an obvious historical chain. Briefly, we shall ask: what were the central features of experience for children in early New England? In what particular ways were children at risk, and how was risk modified or minimized on their behalf? To what extent is there evidence of actual child abuse in that setting (as we have chosen to define the terms)? And how, at last, are we to assess such evidence so as to yield a more general conclusion to our inquiry?

III

Family life in early New England has been the subject of considerable study in recent years; hence we have a relatively firm backdrop to use as our starting point. The children of this particular time and place were, first of all, *numerous*. They arrived at an average rate of 8–10 per married couple, and at any given time they were likely to comprise fully half the population (if we use the term children to mean anyone between the ages 0–16). Then as now, their experience was centered in families—although then, rather *more* than now, they moved easily and informally into activities beyond the home. They faced considerable threats to life and health, though popular lore has exaggerated the sum total of their mortality. Recent investigations show that some 70 to 80 percent of all infants safely born could expect to survive to adulthood.[11]

Unlike the pattern of our own day, children in early New England did not form a sharply delineated age group. For most, schooling was a sometime thing (limited to a few years or seasons), and there was little else in the organized life of the community to set them apart. They joined their elders in the work of fields and farm, in the pleasures of the hearth, in the social round of village and neighborhood, in the devotions of the local church. They were, in effect, "apprenticed" for adult life from an early age, and their growth proceeded a long a relatively smooth and seamless track. It has become fashionable to characterize pre-modern children as "miniature adults," and—for little New Englanders at least—this term does come near the mark.[12]

Was there less feeling invested in them than in children nowadays? Historical evidence of feeling seems always to come in short supply, but what we have suggests strong and enduring ties. A country poet (and wife of a farmer and local magistrate) wrote movingly of the deaths of various infant grandchildren: "No sooner came but gone, and fallen asleep/Acquaintance short, yet parting caused us weep. . . ." A village yeoman remembered that "when his child was sick, and like to die, he ran barefoot and barelegged, and with tears" through the night to find assistance. A local goodwife described the funeral of her cousin, and the plight of "a poor babe" left motherless: "I then did say to some of my friends that . . . I could be very willing to take my cousin's little one and nurse it. . . ." Similar expressions could be assembled in large amounts, but the drift is immediately clear.[13]

Of course, the presence in individual instances of tender feeling toward children would not by itself guarantee their immunity from harm—or even from "abuse." Hence we must try to envision them in the social situations which chiefly framed their lives. This aim directs us, in turn, to one particular category of source materials: the records of the legal system. New England courts were notoriously active in all phases of social experience. There, if anywhere, we might expect to find a mirror for the experience of individual children.

Our expectations are not ill-founded. From time to time the records perserve a case such as the following: "Alexander Edwards complains against Thomas Merrick in an action of the case, for abusing his child named Samuel Edwards, being about 5 or 6 years old, the 14th of April last . . . The witnesses: John Mathews, Nathaniel Bliss. They proved 3 batteries besides villifying words [such] as 'hang him; better kill him than he kill my child'."[14] Such notations show how young children might become implicated in neighborhood squabbles. But they are not very numerous (when the records are considered as a whole), nor, in the average case, do they indicate much real harm to the young "victims."

There were, in addition, particular cases of violence used (or threatened) against children in their own homes. These are obviously pertinent to our inquiry, and they deserve a more extended review. Some—perhaps the largest number—involved an assault on a young bond-servant by his (or her) master. A tragic instance came to light in 1655 in the Plymouth Colony. A boy of thirteen was found dead, his body discolored, "his skin broken in diverse places . . . [and] all his back [covered] with stripes given him by his master." Further investigation disclosed a sequence of antecedent cruelties: frequent and unreasonable whippings, demands for heavy work far beyond the boy's capacity, periodic deprivation of food and clothing. The master was eventually tried and convicted on a charge of manslaughter, and sentenced to be "burned in the hand, . . . and . . . all his goods confiscated."[15]

This was an extreme case. In other instances servants had received "hard usage" which, however, stopped short of a fatal result. They ranged from one child allegedly assaulted by her master's own children to another "found . . . beaten black and blue, with many marks on her body, so that some doctors despaired of her life."[16] The issue was usually *physical* beating, and the point for the court to decide was whether such practice had exceeded "reasonable" bounds. We must understand that the right (and duty) to administer physical "correction" underlay many forms of authority in this culture. Parents would "chastize" their own children in this way—no less than their servants. And the courts of law regularly prescribed whipping as punishment for adult offenders of many kinds.

Occasionally the courts heard charges of abusive conduct toward children by step-parents. Thus, for example: "Henry Merry, of Woburn, . . . [was called] to answer for the cruel beating [of] John Wallis, his wife's child, about 4 years old."[17] But these cases amount to a mere handful overall. Last—and apparently least in a quantitative sense—are the cases that involved parents and their own children. Here is a modest example, drawn from the records of the Essex County Court in Massachusetts: "Michael Emerson, for cruel and excessive beating of his daughter with a flail swingle and for kicking her, was fined and bound to good behavior."[18] And here is a much more affecting instance, as described in John Winthrop's journal:

> A cooper's wife of Hingham, having been long in a sad, melancholy distemper near to frenzy, and having formerly attempted to drown her child, but prevented by God's gracious providence, did now again take an opportunity, being alone, to carry her child, aged three years, to a creek near her house, and stripping it of the clothes, threw it into the water and mud. But the tide being low, the little child scrambled out, and taking up its clothes, came to its mother who was set down not far off. She carried the child again, and threw it in so far as it could not get out; but then it pleased God that a young man coming that way saved it. She would give no other reason

for it, but that she did it to save it from misery, and withal that
she was assured she had sinned against the Holy Ghost. . . .
Thus doth Satan work by the advantage of our infirmities.[19]

A similar case (but with fatal result) is reported elsewhere in the
same document. In 1638 one Dorothy Talbie "was hanged at
Boston for murdering her own daughter, a child of three years
old." She had previously attempted the lives of her husband, her
other children, and her own self. According to Winthrop she was
moved to all this "through melancholy and spiritual delusions"—
and, in the killing of her daughter, was led by impulses from the
Devil which she mistook "as revelations from God."[20]

These two cases require special comment. The principals seem
to have been profoundly disturbed—"distempered" and "deluded"
by the lights of their own time, and probably "psychotic" by ours.
No other instance of plainly murderous intent, on the part of a
parent toward a child, appears within the entire range of early
New England source materials. Perhaps, therefore, the actions of
Dorothy Talbie and the Hingham cooper's wife should be assigned to a small residual category of quite extreme psychopathol-
ogy. The latter *may* be transhistorical, but would, in any culture,
account for only an outer fringe of abusive behavior. Most such
behavior in our own society cannot be associated with full-blown
psychosis.

The total of the evidence to this point seems modest indeed.
The occasional cases in which children suffered abuse from "masters" or neighbors do not fall within our definitional boundaries
as outlined above. And the instances involving parents or step-
parents are very few in number and highly exceptional in circumstance. Yet there remains one further possibility: perhaps some
portion of parental abuse could have been disguised as accidental
injury (evidently a common occurrence today). In this connection
we may profitably examine the records of inquest in cases of
sudden death (where cause was not immediately ascertainable).
Local magistrates regularly empanelled special juries to investigate such cases, at least a portion of which involved children. The
resultant reports were entered in the files of the courts; here is a
typical specimen:

We whose names are underwritten, being warned to serve on a
jury of inquest, have made search on the corpse of Thomas
Evans, aged about ten years. [We] do find that his father and

> he coming from the brook commonly called Griffin's Brook
> with a barrel of water in a horse cart, and he sitting in the cart
> behind the barrel, as they were coming up the hill the mare
> gave a stop. And suddenly moving again, the barrel rolled out
> and drove the child before it, and pitched [him] on his fore-
> head. And there [he] lay until his father took it from him. And
> we find that his skull was very much broken, which was the
> cause of his death.[21]

Other young victims of fatal accidents included "a child of between 3 and 4 years old" crushed by a log which "rolled down"
from a sled he was "endeavoring to get up upon;" a boy of four run
over by his father's cart ("no person knowing it, it being in the
dark of the evening"); still another boy thrown from a cart, "and
the wheel, as we conceive, went over his head;" and a child "found
dead in the brook . . . drowned . . . through its own weakness,
without the hand of any other person being any occasion or cause
thereof."[22]

The court reports on such cases can be supplemented by occasional comments in personal documents. Thus Winthrop's journal describes "a very sad accident at Weymouth" in which a five-
year-old was fatally wounded while playing with his father's gun:

> He . . . took it, and laid it upon a stool, as he had seen his
> father do, and pulled up the cock (the spring being weak), and
> put down the hammer, [and] then went to the other end and
> blowed in the mouth of the piece, as he had seen his father also
> do; and with that stirring [and] the piece being charged, it
> went off and shot the child into the mouth and through his
> head.[23]

In another instance an infant died in a fire started by an older
sibling. (The latter had inadvertently "burned a clot, and fearing
its mother should see it, thrust it into a haystack by the door, . . .
the fire not being quite out. . . .")[24] The particulars on this list
make painful reading, yet they convey no hint of parental abuse—
and little enough of parental neglect. Over-running by carts,
drownings, housefires, gun mishaps: these form an expectable
range of accidental outcomes in a pre-modern village setting.

The judicial inquests into childhood death belonged to a larger
pattern of oversight and protection. Local officials stood ready to
intervene whenever the care of particular children seemed imperiled by the wants and failings of their parents. In March 1680, for

example, the selectmen of Northampton, Massachusetts, obtained a court order to remove three children from their own home and place them "to service as apprentices" in other households. The parents, it seems, were "in a very low condition . . . and [lacked] things necessary for supplies and bringing up of [their] children." The same court imposed similar orders on a second family where the children's "education" was neglected—"it appearing that the father of them is very vicious and rather [is] learning them irreligion than any good literature."[25] These measures (and a host of others like them) reflected official concern with the bodily and spiritual "estate" of young persons thought to be at risk.

In most instances court action was itself the result of intervention by others with personal knowledge of the key circumstances. For example: an elderly Massachusetts woman brought suit against her son-in-law for neglecting his children (who were also *her* grandchildren), and neighbors volunteered testimony as to their "suffering condition."[26] Probate records supply convergent evidence in large quantities. A man writing a will would create specific lines of responsibility for the welfare of his under-age children. Thus, in one case: "My will and mind is that if my wife Edna Bailey marries again, and her husband prove unloving to the child . . . I give power to my brother James Bailey and [to] Michael Hopkinson, with my wife's consent, to remove the child, with his portion [i.e. of inheritance] from him, and so to dispose of it for the best behoof of the child." And again, in another case: "I appoint these five Christian dear loving friends . . . to be my executors and administrators of this my last will and testament, as also to be the overseers of my wife and children in a friendly Christian way toward them, and [I direct] that you five should take the advice of our [church] elders."[27] Here, in sum, was a web of responsibility for children extending from parents and step-parents to relatives, friends, estate executors, religious leaders—and, as the final resort, civil authorities.

The details of all this demand our attention because the issue at hand is a tricky one. We are trying to understand *an absence of evidence* (e.g. of behavior which might qualify as child abuse). And the outcome depends on careful evaluation of other material with implicitly related meaning; in short, we need to use what *is* in the record to interpret what *does not* appear there. In such cases one must always recognize the possibility that the designated behavior was present, and perhaps even widely so, in reality, but without leaving traces for the historian to follow. (Perhaps it was taken so much for granted by contemporaries that it would not attract notice—and thus not enter the historical record. Or perhaps the evidence it generated has not survived the passage of time.) In the current instance, however, this possibility cannot be fitted to the known facts. Consider the following, by way of summary of the substantive materials reviewed in the preceding pages.

(1) Had individual children suffered severe abuse at the hands of their parents in early New England, other adults would have been disposed to respond. The culture, in general, seems to have sponsored a solicitous attitude toward the young. We know moreover, that in the (relatively infrequent) cases where *servant*-children were brutalized by masters, public sympathy was directly engaged. Hence, we may reasonably infer a *motive* to prevent child abuse.

(2) The powers of local magistrates, and the procedures of the courts, would readily have invited action against child abuse. In our own time such action is sometimes inhibited by a distaste for intruding into another family's affairs; indeed the abuse itself may be long concealed within the private space to which all families feel entitled. But in early New England little value attached to personal and domestic privacy. Elected officials were empowered to oversee family life, no less than other aspects of life. Neighbors, in fact, accepted informal oversight and responsibility for one another as part of their "Christian duty." In short, there were *means* to discover child abuse (if it should occur) and to take the necessary remedial steps.

(3) Had the courts and/or other official bodies been obliged to deal with child abuse, there would now be sufficient *evidence* to alert us to the fact. The early New Englanders kept elaborate records of their various doings—records which have, for the most part, been preserved with loving care by successive generations of their descendants.

These considerations of motive, means, and evidence allow us to interpret the absence of records of child abuse as an absence of the behavior itself. Our "case study" of early New England is now concluded. And it throws a long shadow of doubt on the "consensus view" of this subject.

IV

While we have no other historical "cases" to set alongside the material from early New England, there is interesting and convergent evidence in recent work in anthropology. There, too, the question of child abuse can be put to pre-modern conditions; the results so far fall into two parts. On the one hand, anthropologists feel constrained to underscore the range of culturally normative standards in child-rearing; thus one society's abuse may be another's common practice. But on the other, the "classic child abuse complex" of the modern West (as one writer has called it) is not found in traditional non-Western societies. (This "complex" conforms roughly to the definition formulated earlier in our discussion.) Moreover, where the old order breaks down in the face of incipient modernization (e.g. urbanization, industrial development, etc.), child abuse begins to appear in its "classic" form.[28]

These findings from a variety of cultures around the contemporary world return us by a natural route to the evolution of our own society. We are confronted, finally, with an historical *conundrum* of large and unsettling proportions. If child abuse is far more prevalent now than two or three centuries ago, the question arises—*why*? What factors, which processes, unraveling through the course of time, account for the change? It is obviously impossible to canvass all the relevant possibilities here, but perhaps we can trace at least the outline of an answer, if only to invite further thought and discussion.

The scientific literature on present-day child abuse provides a good point of entry. Briefly summarized, this literature presents a three-part model of the key "predisposing influences": (1) environmental ones (social and cultural forces, in general); (2) situational ones (pressure-points within abusing families, in particular); (3) psychological ones (the character of abusing parents as individuals).[29] Each of these can be stretched out along the frame of time and joined to other elements in our historical past.

Environmental interpretations of child abuse invariably highlight a pair of central tendencies in modern American society. One is the fact of endemic violence, and a climate of values which condones violence under at least some circumstances. The other is the condition of social isolation in which many individuals and families find themselves obliged to live. Both these tendencies may be considered as historical growths, the second perhaps more strikingly so than the first. Pre-modern society was, if nothing else, quite fully integrative of the lives of its constituent members. The traditional village setting—for example, in early New England—offered to each person and every family a density of human contact that is hard even to imagine today. The marketplace, the church, the court, the broad spectrum of local routine and custom made a tight web of social experience allowing few possibilities of escape

or exclusion. Pre-modern communities had their share of noncon-
formists, eccentrics, and criminals—but no isolates, no habitual
"strangers." The shape of life from day to day expressed the twin
principles of mutual support and mutual surveillance. It is in just
this regard that the situation of our own abusing parents seems
most sadly deficient. Study after study finds them rootless, friend-
less, virtually unknown even to next-door neighbors.[30] Are they
overwhelmed by real or imagined adversities? There is no one else
to share the load. Do they lash out at the nearest available human
targets? There is no one to see the hand rise, and to stay its swift
descent.

Other features of contemporary society must also be mentioned
here. Unemployment appears to correlate with child abuse (and
with wife-beating as well). And there is reason to think that on-
the-job alienation (boredom, frustration, a sense of "depersonali-
zation") may show a similar link.[31] These *residua* of our modern
economy find no parallel in pre-modern times. Work could be—
often was—hard and very meager in its rewards, but it was not
simply *denied* to some considerable portion of the available hands.
Moreover, there was no equivalent to the numbing routines of
production in the mass—the assembly line, the secretarial pool,
the shapeless ranks of white-collar bureaucracy.

One more part of this environmental matrix is the pace and
power of change itself. According to a recent quantitative study,
"abusing parents" have a disproportionate experience of "crisis,"
as measured on a "social readadjustment scale" of "life change
units." The details need not concern us; suffice it to say that the
scale in question includes many items hard to associate with a
pre-modern setting (e.g. mortgage foreclosure, job change, wife
entering the labor force, retirement, "business readjustment," divorce).[32] It is arguable, in short, that change is now a more frequent and disruptive life-presence than was the case a few generations ago. It also seems likely that for many of us the capacity to
accept and absorb change is less. We have no clear equivalent to
the "providential" world-view of our forebears—their belief that
all things, no matter how surprising and inscrutable, must be
attributed to God's overarching will.

We turn next to the "situational" factors most often adduced
with respect to child abuse. "In abusing families," notes one
authority, "there is a constant competition over who will be taken
care of." The parents themselves enter this competition with great
intensity; they seek, in effect, a "role reversal" in which the chil-

dren will become their caretakers. Life in such households is characterized by mutual overinvolvement, amounting in some cases to "emotional fusing"; there is a pervasive quality of "stuck-togetherness."[33] Described in more formal terms, these people are poorly differentiated from one another; hence they fall easily into scapegoating, schemes of manipulation, and all manner of misunderstanding.

Here situational elements merge with psychological ones—the third part of our explanatory model—and the personality of the abusing parent comes fully into view. Again, there is a clear thematic center common to all recent studies: "low self-esteem," and feelings of emptiness, worthlessness, helplessness.[34] Viewed from the standpoint of current psychoanalytic psychology, child abusers suffer from severe deficits in basic "narcissism."[35] Their inner lives are organized around archaic (and largely unconscious) "grandiose fantasies" for themselves and literally heroic (or "idealized") expectations of others. They are unable to make consistent self/other distinctions; indeed, they characteristically approach others as extensions of their own selves.

The connection to child abuse goes roughly as follows. The parent is unable to recognize the child as a separate individual with needs and aims in his/her own right. When the child, inevitably and appropriately, asserts his independent self, the parent is surprised, disoriented, even infuriated. So it is that a mere cry at an inopportune time (inopportune, that is, from the parent's standpoint) can be experienced as a deep flaw in a narcissistically perceived world—indeed as an outright affront to that world. Like all such affronts it threatens inner "fragmentation" (for the parent), and the likely response is that special form of uncompromising anger known to clinicians as "narcissistic rage."[36]

This excursion into clinical theory, brief and oversimplified though it is, will have to serve as the basis for one final group of historical speculations. We have tried to underscore the particular "family constellation" and the (related) psychological set which appear most conducive to child abuse. But, again, we wish to explore the contrast between *then* and *now*. That contrast can be followed through several different, though apparently overlapping, historical dimensions.

(1) The *demographic* dimension: We start from the fact that families in early America were characteristically *large*. As previously noted, a range of 8–10 children per married couple was typical. We may infer that in households of this size and shape the

emotional exchange between parents and children was necessarily diffused. For one thing, a certain portion of "parenting" was carried on by older children (vis-à-vis younger ones); for another, there were kin and neighbors nearby who would enter the family circle on a fairly regular basis. By contrast, the modern American family includes significantly fewer persons, and its style of interaction seems far more intense. The element of psychic involvement—specifically including "narcissistic" involvement—is correspondingly deepened.[37]

(2) The *structural* dimension: It hardly needs saying that the place of children in the productive sector of family life has greatly diminished over the past several generations. Time was when the young performed valuable service as part of a "working household" (e.g. the farm family of early New England, described above); now, of course, they constitute a substantial drain on family resources.[38] This makes them *vulnerable* in ways both new and profound.

(3) The *normative* dimension: In pre-modern times the fate of each individual was related, by prevalent custom and belief, to forces well beyond his human surroundings: to inherited status in a traditional community, to the "accidents" of Nature, and above all to the ultimate purposes of God. But beginning in the nineteenth century this view was replaced by a new ethos of "individualism." From henceforth personal destiny was seen, in part, as something self-determined—and, in equal part, as dependent on one's immediate family. Parents, in particular, were charged with profound responsibility for the life and prospects of their children; little people could be "set on a true course," or ruined for life, by the influence of their home environment. (Conversely, parents near the end of their lives could be vindicated, or destroyed, by the performance of their grown children.) As part of this new belief-set there developed the implicit notion that young children are somehow the *property* of their parents—part of the larger armamentarium with which a family faces the "outside world." This idea of children-as-property has been deeply consequential in several directions; not least, it has validated new concepts of parental "rights" and prerogatives.

Moreover, ours is a culture which accords an almost unique significance to measures of individual "success" and "failure." These measures are, in the first instance, economic and occupational ones, but they reach out in other directions as well (e.g. good looks, good "personality," etc.). There is room enough here for the cultivation of self-doubt—and the lessening of healthy "narcissism"—in persons of every type and background. Once again, it was not always so. When individuals measured their worth in terms of prescribed position in a given community and presumed status in the eyes of God, their experience of "self" was in some ways easier.

V

The thread of this discussion, now at an end, has been long and somewhat tortuous. We began by considering the current consensus-view that the history of childhood—and, more particularly, of child abuse—is essentially "progressive." We found reason to question this view and the scattershot evidence on which it depends. Next we developed a "case study" of our subject, from materials on one particular historical setting. The outcome cast further doubt on progressive interpretations. We noticed, in passing, the convergent findings of anthropologists concerned with pre-modern culture. And we concluded with some broad hypotheses as to why child abuse may have become more prevalent, and more severe, within the past two or three centuries.

There is danger always, when refuting one historical judgment, of rushing to the opposite extreme. We must not resurrect an older, plainly romanticized notion of the past with its down-on-the-farm sentimentality about families, neighborhoods, and life in general. Most childhoods in pre-modern society knew their own forms of severity. But they seem *not* to have known the particular sufferings which the term "child abuse" now calls so vividly and painfully to mind.

The pain is, of course, a crucial point. Open any casebook, examine any photographic file on battered children—and the impulse to look away is overwhelming. Dr. Kempe has noted the contorted efforts of physicians, in the face of "obvious" signs, to "obliterate such suspicions [of child abuse] from their minds." If we are not careful, history may lend itself to similar stratagems. It is easier to think that this problem, however terrible today, must have been far worse in the past. In such historical comparisons we seek a measure of comfort—not to say, a defense against truth.

And there are other truths with which this one makes a disconcerting fit. The society that finds within itself the malignancy of child abuse is the same society which has, over a few short generations, experienced assassinations of public figures, racial repressions, bitterly divisive wars, and ever-rising levels of street violence.

These are grievous wounds, and they call for massive projects of healing. Historians are not healers, but they can and must contribute to the process of truth-seeking, of diagnosis, of *understanding*—from which alone the healing may come.

NOTES

1. C. Henry Kempe, F. N. Silverman, B. F. Steele, William Droegmueller, and Henry K. Silver, "The Battered Child Syndrome," in *Journal of the American Medical Association,* CLXXXI (1962), 17–24.

2. The legal and legislative aspects of all this are summarized in Monrad G. Paulsen, "The Law and Abused Children," in *The Battered Child,* Ray E. Helfer and C. Henry Kempe, eds., 2nd ed. (Chicago, 1974), 153–78. See also "Appendix B: A Summary of Child Abuse Legislation, 1973," in *ibid.,* 203–27; Vincent DeFrancis, *Child Abuse Legislation: Analysis and Study of Mandatory Reporting Laws in the United States* (Denver, 1966); Joseph J. Costa and Gordon K. Nelson, *Child Abuse and Neglect: Legislation, Reporting, and Prevention* (Lexington, Mass., 1978); Jean M. Giovannoni and Rosina M. Becarra, *Defining Child Abuse* (New York, 1979); and Barbara J. Nelson, *Making An Issue of Child Abuse: Political Agenda-Setting for Social Problems* (Chicago, 1984).

3. The history of medicine, in relation to child abuse, is summarized in David Bakan, *Slaughter of the Innocents* (San Francisco, 1971), 44–54. On the development, specifically, of pediatric radiology, see F. N. Silverman, "Radiologic Aspects of the Battered-Child Syndrome," in Helfer and Kempe, eds., *The Battered Child,* 2nd ed., 41–60.

4. The "Mary Ellen case" is described in Vincent J. Fontana, *The Maltreated Child* (Springfield, Ill., 1964), 9. Developments subsequent to that case are described and discussed in Samuel X. Radbill, "Children in a World of Violence: A History of Child Abuse," in C. Henry Kempe and Ray E. Helfer, eds., *The Battered Child,* 3rd ed. (Chicago, 1980), 3–20; Mason P. Thomas, "Child Abuse and Neglect: Historical Overview, Legal Matrix, and Social Perspective," in *North Carolina Law Review* L (1972), 293–349; and Catherine J. Ross, "The Lessons of the Past: Defining and Controlling Child Abuse in the United States," in George Gerbner, Catherine J. Ross, and Edward Zigler, eds., *Child Abuse: An Agenda for Action* (New York, 1980), 63–81. For an excellent treatment of parallel developments in England, see George R. Behlmer, *Child Abuse and Moral Reform in England 1870–1908* (Stanford, Calif., 1982); on the situation in France, see Rachel Fuchs, "Crimes Against Children in Nineteenth-Century France: Child Abuse," in *Law and Human Behavior.* VI (1983), 237–259. Important new work on child abuse in American history will appear shortly in Elizabeth H. Pleck, *Domestic Tyranny: The Process of Reform Against Family Violence* (New York, 1986).

5. Fontana, *The Maltreated Child* 3; Radbill, "Children in a World of Violence," 3; Bakan, *Slaughter of the Innocents,* 25.

6. Lawrence Stone, *The Family, Sex, and Marriage in England, 1500–1800* (New York, 1977), 95; Lloyd deMause, "The Evolution of Childhood," in de-Mause, ed., *The History of Childhood* (New York, 1974), 40.

7. This viewpoint is found, for example, in Stone, *The Family, Sex, and Marriage in England, 1500–1800*, in deMause, "The Evolution of Childhood," and in Edward Shorter, *The Making of the Modern Family* (New York, 1975).

8. deMause, "The Evolution of Childhood," 51–54 and *passim*.

9. Bruno Bettelheim, *The Uses of Enchantment: The Meaning and Importance of Fairy Tales* (New York, 1975).

10. For a discussion of problems in the definition of child abuse see David G. Gil, *Violence Against Children: Physical Child Abuse in the United States* (Cambridge, Mass., 1970), 5–8. The same problems are addressed from a legal standpoint in Giovannoni and Becarra, *Defining Child Abuse*, and from an anthropological standpoint in Jill Korbin, "The Cross-Cultural Context of Child Abuse and Neglect," in Kempe and Helfer, eds., *The Battered Child*, 3rd ed., 21–35.

11. On childhood in early New England, see John Demos, *A Little Commonwealth: Family Life in Plymouth Colony* (New York, 1970), chs. 4, 9, 10; Edmund S. Morgan, *The Puritan Family: Religion and Domestic Relations in Seventeenth-Century New England*, rev. ed. (New York, 1966), chs. 3–5; Joseph Illick, "Child-Rearing in Seventeenth-Century England and America," in deMause, ed., *The History of Childhood*, 303–50; and Philip J. Greven, Jr., *The Protestant Temperament: Patterns of Child-Rearing, Religious Experience, and the Self in Early America* (New York, 1977).

12. The image of children as "little adults" was investigated at length by the French cultural historian Philippe Ariès in his path-breaking study published in English under the title *Centuries of Childhood*, Robert Baldick, trans. (New York, 1962). See also J. H. Plumb, "The New World of Children in Eighteenth-Century England," in *Past and Present*, LXVII (1975), 70–85, and Demos, *A Little Commonwealth*, ch. 9. For a somewhat different viewpoint, see Ross Beales, "In Search of the Historical Child," in *American Quarterly*, XXVII (1975), 379–98.

13. Anne Bradstreet, "On my Dear Grandchild, Simon Bradstreet, Who Died on 16 November 1669, Being But a Month and One Day Old," in *The Works of Anne Bradstreet*, Jeannine Hensley, ed. (Cambridge, Mass., 1967), 237; "Examination of Hugh Parsons, of Springfield, on a Charge of Witchcraft . . . Before Mr. William Pynchon, at Springfield, 1951," printed in Samuel G. Drake, *Annals of Witchcraft in New England* (New York, 1869), 239; "Bethiah Lothrop's Statement," given to the Essex County Quarterly Court, 27 June 1676, printed in *The Probate Records of Essex County* (Salem, Mass., 1916–20), III, 28. In general, such evidence of emotional experience is located in the depositional statements presented in numerous civil actions before local courts. Since this is largely manuscript material, it is not well-known even to professional historians.

14. From the records of the Hampshire County (Mass.) Court, 21 April 1648, as kept by William and John Pynchon, and published in James M. Smith, ed., *Colonial Justice in Western Massachusetts* (Cambridge, Mass., 1961), 216.

15. Nathaniel B. Shurtleff and David Pulsifer, eds., *Records of the Colony of New Plymouth, in New England* 6 vols. (1855–61), III, 71–73.

16. Case of William Holdred vs. John Ilsly, Salem Quarterly Court, 17 October 1666, published in *Records and Files of the Quarterly Courts of Essex*

County, Massachusetts, 8 vols. (1911–21), III, 365; petition of Henry and Jane Stacy, Salem Quarterly Court, 29 June 1680, in *ibid.,* VII, 421.

17. Case heard by the Middlesex County (Mass.) Court, 18 April 1662, in Middlesex County Court Reocrds, vol. I, leaf 105 (manuscript volume, Clerk's Office, Middlesex County Courthouse, East Cambrige, Mass.).

18. Case heard by the Ipswich Quarterly Court, 23 May 1676, published in *Records and Files of the Quarterly Courts of Essex County, Massachusetts,* VI, 141.

19. *Winthrop's Journal,* James Kendall Hosmer, ed. (New York, 1908), II, 60.

20. *Ibid.,* I, 282.

21. Records of the County Court of Hartford (Conn.), 29 October 1683, in Hartford Probate Records, vol, IV, leaf 74 (manuscript volume, Connecticut State Library, Hartford).

22. Records of the Hampshire County (Mass.) Court, as kept by William and John Pynchon, 27 December 1674, in Smith, ed., *Colonial Justice in Western Massachusetts,* 282; records of the Hampshire County (Mass.) Court, 25 September 1660, in *ibid.,* 248; records of the Essex County (Mass.) Court, 29 September 1676, in *Records and Files of the Quarterly Courts of Essex County, Massachusetts,* VI, 234; records of the Hampshire County (Mass.) Court, in Smith, ed., *Colonial Justice in Western Massachusetts,* 245–46.

23. *Winthrop's Journal,* Hosmer, ed., II, 72.

24. *Ibid.,* II, 30.

25. Records of the Hampshire County (Mass.) Court, 30 March 1680, in Hampshire County (Mass.) Probate Records, vol. I, leaf 205 (manscript volume, Registry of Probates, Northampton, Mass.).

26. Mrs. Elizabeth King vs. John Blano, Salem Quarterly Court, 29 June 1677, in *Records and Files of the Quarterly Courts of Essex County, Massachusetts,* VI, 299–300. See also some later proceedings in this case, Salem Quarterly Court, 12 November 1677, in *ibid.,* VI, 359–61.

27. Will of Richard Bailey, of Rowley, (Mass.), 15 February 1647, in *The Probate Records of Essex County,* I, 92; will of John Lowell, of Newbury (Mass.), 29 June 1647, in *ibid.,* I, 67.

28. See Jill Korbin, "Anthropological Contributions to the Study of Child Abuse," in *Child Abuse and Neglect: The International Journal,* I (1977), 7–24. Korbin has also edited an excellent anthology of essays on the treatment (and maltreatment) of children in a variety of cultures around the world: *Child Abuse and Neglect: Cross-Cultural Perspectives* (Berkeley, Calif., 1981). In many respects the settings described there differ drastically from one another. However, they all support the general proposition stated in a Foreword by Robert B. Edgerton: "Child abuse . . . has become a serious social problem in the United States and in some other industrialized societies, yet it occurs infrequently or not at all in many of the world's [non-industrialized] societies" (ix).

29. On "causal models" in the study of child abuse, a fine summary can be found in R. D. Parke and C. W. Collmer, "Child Abuse: An Interdisciplinary Analysis," in E. M. Heatherington, ed., *Review of Child Development Research,* V (Chicago, 1975), 509–90. See also Blair Justice and Rita Justice, *The Abusing Family* (New York, 1976), 37–54, and Mildred Daley Pagelow, *Family Violence* (New York, 1984), 74–143.

30. See, for example, Leontine Young, *Wednesday's Children: A Study of Child Neglect and Abuse* (New York, 1964), 37ff.; Justice and Justice, *The Abusing Family*, 94; and Brandt F. Steele and Carl B. Pollock, "A Psychiatric Study of Parents Who Abuse Infants and Small Children," in Helfer and Kempe, eds., *The Battered Child*, 2nd ed., 106.

31. Justice and Justice, *The Abusing Family*, 255–57; Leroy H. Pelton, ed., *The Social Context of Child Abuse* (New York, 1981); James Garbarino and Gwen Gilliam, *Understanding Abusive Families* (Lexington, Mass., 1980). The "ecological" approach to the child abuse problem has assumed major importance in recent years, and the literature exemplifying this approach is enormous. For additional authors and titles, see Elizabeth Kemmer, *Violence in the Family: An Annotated Bibliography* (New York, 1984).

32. See Murray A. Strauss, "Stress and Child Abuse," in Kempe and Helfer, eds., *The Battered Child*, 3rd ed. (Chicago, 1980), 86–103; Murray A. Straus, Richard J. Gelles, and Suzanne K. Steinmetz, *Behind Closed Doors: Violence in the American Family* (New York, 1980), esp. ch. 8; Justice and Justice, *The Abusing Family*, 25–34.

33. *Ibid.*, 61ff. See also M. Bowen, "The Use of Family Theory in Clinical Practice," in *Comprehensive Psychiatry*, VII (1966), 345–74; Murray A. Straus, "A General Systems Theory Approach to a Theory of Violence Between Family Members, in *Social Science Information*, XII (1973), 105–25; and Susan K. Steinmetz, *The Cycle of Violence: Assertive, Aggressive, and Abusive Family Interaction* (New York, 1977).

34. A good general survey of the psychiatric literature on child abuse can be found in Brandt Steele, "Psychodynamic Factors in Child Abuse," in Kempe and Helfer, eds., *The Battered Child*, 3rd ed., 49–85. See also Parke and Collmer, "Child Abuse: An Interdisciplinary Analysis," and Gerbner *et al.*, *Child Abuse*, chs. 2–3.

35. The best recent work in the clinical theory of narcissism is that of Heinz Kohut and his colleagues at the Institute of Psychoanalysis in Chicago. See, for example, Kohut, *The Psychology of the Self* (New York, 1970) and *The Restoration of the Self* (New York, 1977).

36. This paragraph is a summary of studies currently in progress, under the direction of Gustavo Lage, M.D., of the Institute of Psychoanalysis in Chicago. I am much indebted to Dr. Lage for various professional communications on the subject.

37. On the issue of "narcissistic" involvement, in modern American families, see Kohut, *The Restoration of the Self*, 269–270. See also John Demos, "Oedipus and America: Notes on the Reception of Psychoanalysis in the United States," in *The Annual of Psychoanalysis*, VI (1978), 23–39.

38. "The total costs of housing, feeding, and clothing one child, as well as educating him or her through high school, now add up to more than $35,000 by very conservative estimates for a family living at a very modest level." See Kenneth Keniston and the Carnegie Council on Children, *All Our Children: The American Family Under Pressure* (New York, 1977). More recent estimates run over $100,000. For a particularly sensitive treatment of all the issues surrounding the "costs" of raising children, see Viviana A. Zelizer, *Pricing the Priceless Child: The Changing Social Value of Children* (New York, 1985).

CHAPTER V

The Rise and Fall of Adolescence

The study of the life course is one of the most intriguing, and most difficult, areas of modern historical scholarship. Clearly, every one of us has a life course of some sort; every one of us undergoes a sequence of movement and change, with a beginning and an end, and with a sense of experiential parts connected to one another and to the whole. Clearly, too, there are certain underlying fixities about all this: we all start small, grow larger and (to some degree) wiser, develop new capacities (e.g. to reproduce ourselves), and finally grow old and die. But within these very broad boundaries the life course allows an almost infinite variety of permutations and combinations. No two life courses are identical, or even close to identical. For genetic reasons, as well as circumstantial ones, each has its own line to trace, its own story to tell.

And there is something in between—something apart from the fixed boundaries, on the one hand, and the idiosyncratic stories on the other. Communities, cultures, and historical epochs impose a measure of regularity, of similarity, on the life course: similarity, that is, among those who share a given cultural/historical setting—but difference from those in other settings. This middle ground is the territory of the historian (and, perhaps, of the anthropologist). What is similar—he will ask—about the life course as experienced by modern-day Americans? And wherein does such similarity make them different from pre-modern Americans, or Samoans, or Eskimos?

My own interest in this territory began with the particular "stage" we call adolescence. As a beginning graduate student I was assigned the following question: "Where did the modern idea of adolescence originate?" The assignment proved interesting and instructive, but it led on inevitably to a further question: "Why did the modern idea of adolescence originate?" This, in turn, was part of my own passage from "intellectual" to "social" history. My original seminar paper (the assignment) was an account of psychologists and educators creating a set of concepts. But no sooner had I finished than I wrote a second paper (in company with a particularly close colleague) treating the social circumstances to which those concepts responded. More recently, I have attempted yet a third version, which incorporates elements from each of the first two (along with much valuable work performed in the interim by other hands).

More than anything else I've done, my writings on adolescence have been challenged and criticized. The "third version" (printed here) is partly a response to the critics; it admits to some prior overstatement but stubbornly reasserts the main lines of my original argument.

The historical study of adolescence may be said now to have entered its own adolescence. Born less than twenty years ago, as a spin-off of childhood history, it has grown and flourished to an extent that surprises even its progenitors; at the same time, it has developed new marks of complexity and internal conflict. To carry the developmental metaphor a little further: the field itself is experiencing something of an "identity crisis."

The first historians of adolescence pursued a singular track—singular because it seemed to disappear scarcely a century into the past.[1] Adolescence, they concluded, was largely a creation of recent times. Or at least the *idea* of adolescence was such a creation—and, to be sure, their research was directed more to ideas than to experience. The gist of their argument went as follows. The modern currency of the idea of adolescence owes much to the efforts of one man, Professor Granville Stanley Hall. Hall's work is now largely outdated and forgotten; if he is remembered at all, it is probably for his role as host to Sigmund Freud during Freud's one and only visit to America. (Freud came in 1909 to deliver a series of lectures at Clark University in Worcester, Massachusetts. Hall was the university's president and the person most responsible for making the invitation.) In his own time Hall was a considerable eminence among American psychologists, albeit a controversial one; indeed his reputation extended well beyond academic circles, for he was among other things a remarkably successful popularizer of psychological ideas.[2] Hall's *magnum opus* was a vast two-volume study, published in 1904, of *Adolescence: Its Psychology, and Its Relations to Anthropology, Sex, Crime, Religion, and Education*.[3] The title sounds a bit eccentric—and the book reads rather strangely—today; but at the time it was a blockbuster. Indeed, the term "adolescence," and the ideas which Hall and his disciples subsumed under the term, achieved a kind of instant celebrity. It was almost as if Hall's account of adolescence had filled a gap, or a need, in social experience—so widely and eagerly was it taken up. To be sure, the word itself was not new; and it does have a certifiable Latin root. However, there can be little doubt of its special salience in and for the twentieth century: whether measured quantitatively as a matter of popular usage, or qualitatively for its distinctive timbre and resonance, adolescence looms far larger now than in previous eras.

Such, to repeat, was the first historical view of this subject. Apart from the evidence of psychological theory-building (Hall *et al.*), it owed much to recent research in several of the social sciences. Anthropologists, for example, had frequently noticed an absence of special concern for adolescence in pre-modern cultures around the world. In fact, as far back as the 1920s Margaret Mead had designed her classic study *Coming of Age in Samoa* as a more or less direct refutation of Hall's "saltatory" theory of adolescent experience.[4] Hall believed that human development everywhere included a period of *sturm und drang* between childhood and adulthood; Mead sought to show that *sturm und drang* was culturally conditioned—indeed was quite specific to developmental experience in modern Europe and North America. Meanwhile, sociologists discovered a similar pattern—i.e. less *versus* more highlighting of adolescence—in rural as contrasted with urban populations even within the United States.[5] These contrasts seemed to lend themselves quite readily to historical adaptation, with "rural" and "pre-modern" standing for the further past (most Western countries before about 1800), and "modern"/"urban" designating the recent past up to and including our own time.

As is the way with all scholarship, this view was soon challenged by new evidence and "revisionist" interpretations. The pre-modern histories of England, continental Europe, and the Americas showed, on close inspection, the familiar forms of adolescent self-consciousness and self-expression: youth groups and cults, rebelliousness against adult norms, age-specific patterns of innerlife preoccupation.[6] G. Stanley Hall was correct, after all: adolescence *was* a virtual "universal" (at least in the history of the West). *Plus ça change, plus c'est la même chose.* To some extent, this revisionism expressed a shift of research strategy. Ironically, Hall and his fellow theorists were played down: experience, rather than ideas, became the center of investigation. Hence it mattered little that pre-modern populations lacked our own terms and concepts to identify a particular stage of development. And it mattered much that they felt, and behaved, roughly as we do at the same stage.

Thus have the lines been fairly drawn between two modes of historical interpretation. And, as often happens in such debates, a wise umpire—a wise *reader*—may expect to find the "truth" somewhere in between.[7] Perhaps the first view erred by way of overstatement, and perhaps the second erred by way of over-reaction. History always blends change and continuity, the "universal" and the transitory; the trick is to find the balance in specific cases. But the balance in *this* case seems particularly delicate. For one thing, the evidence is necessarily scattered and fragmentary; for another, historians are not as yet practiced in dealing with such broad, yet personal, subject-categories. It may be useful, therefore, to borrow some definitional markers that have proved their worth elsewhere in the social sciences.

Let us regard this investigation as having three component strands—and as requiring three different interpretive perspectives. Adolescence is, first, a matter of *biology*—that is, of intrinsic developmental process in a strictly physiologic sense. It is also a matter of *psychology*, insofar as it involves the resolution of internal issues around oneself and one's "significant others." Finally, it is a matter of *culture*: here is the point at which ideals, values, norms enter in. There is a sense in which the three aspects comprise a spectrum, from the most nearly universal and unchanging—that is, the biological ones—to the most variable and flexible—that is, the cultural ones. (Psychology seems to lie somewhere in between.)

To be sure, even biology is not entirely impervious to altered historical circumstances. Thus there is reason to think that the attainment of full physical stature comes at a considerably earlier age now than in pre-modern times. According to one estimate young men in early nineteenth-century America did not reach their final adult height until they were about twenty-five years old, whereas most of their contemporaries today do so by or before age twenty.[8] There is a similar point to be made on the timing of menarche in girls. Medical evidence since the middle of the last century suggests that menstruation has been starting nearly one year earlier with each succeeding generation—falling, that is, from an average of about fifteen in 1850 to almost twelve nowadays.[9]

These trends are clearly of interest; they should not, however, be *over*-estimated. No known culture, no historical era, has seen maturation delayed into (the average person's) fourth or fifth decade, or advanced much into the "pre-teen" years. Indeed the age-range of twelve–fifteen (for menarche) and sixteen–twenty-five (for attaining adult height) may well represent outside limits, biologically speaking, for these important developmental benchmarks. The historical period from roughly 1850 to the present appears to have brought unprecedented improvements in nutrition, in health care, and in other factors related to physical maturation—unprecedented and unsustainable in the future. New studies show a leveling, within the past decade or so, of the trend-line for age at menarche; and fragmentary evidence on the same question, from pre-modern times, implies a level pattern there as well. Hence the post-1850 decline may have been atypical and time-limited.

The biological strand of this history is, then, rather quickly played out. Psychology and culture, by contrast, are the heart of the matter. And, as between these two, the second is necessarily our baseline. In all of what follows we lead with the evidence of culture—of values and feeling, of material circumstance, of pattern and "structure" in interpersonal experience. For here the historical record is fullest by far. But psychology is frequently evident *through* culture, and constitutes in any case a key part of the agenda in all life-course studies. Adolescence is today simultaneously a social category and an intrapsychic "stage"? Was it also this way in the past?

I

The colonial period of American history is an appropriate unit for beginning inquiry. It was not, of course, a "unit" in this respect or in any other: differences of both time (e.g. the early seventeenth as contrasted with the late eighteenth century) and space (the Northern *versus* the Middle *versus* the Southern colonies) created their own effects, especially at the level of local and individual experience. Still, viewed from hundreds of years later on, certain general tendencies do come clear.

The single, most widely salient fact about society in this period was its pre-modern, pre-industrial character. The population was distributed through hundreds of village-communities (or, in some regions, more isolated farms and plantations). For at least 90 percent of these "colonists" day-to-day life was shaped by the requirements of small-scale agriculture—and the life, specifically, of *young* people reflected patterns typical of pre-modern farm-youth everywhere. It is clear, for example, that children began from an early age to participate in the productive work of their individual households. Seven- or eight-year-old boys and girls would assume simple responsibilities in the care of domestic animals, in gardening, and in such household tasks as spinning, candle-making, and food-preparation. As they grew older, the scope of these responsibilities widened. Presumably, the details of the process were as variable as the individuals and families directly involved, but its larger meaning for our purposes seems clear: children moved toward maturity along a smoothly surfaced path. Their introduction to adult roles began early; culturally appropriate adult models were present, and visible, from the start. Under these conditions a male child was, in effect, a miniature version of his farmer-father; likewise a female one, in relation to her mother.[10]

This is not to say that children appeared, to themselves or to others, equal in capacity to their elders. On the contrary, they were seen as being inferior—in intelligence, in experience, in practical skills, in physical (and moral) strength. Thus, for example, were certain tasks expressly considered children's work, while others were reserved for adults. And thus, too, were expectations of religious experience adjusted to age: church doctrine declared children incapable of fully apprehending "the mysteries of salvation," and catechism classes set them formally apart from the rest of the congregational community. Still, *all* such forms of inferiority implied difference in scale, not in kind—and presumed enlargement, not change, as the key to youthful development.[11]

The key to this pattern was, as noted, the economic organization of the household; but other factors—for example, demographic ones—also played in. Families were very large, by our standards: eight to ten children per married couple, born over a period of as much as twenty years. This meant an implicit blurring of intergenerational lines. Many children could envision their own progress toward adulthood by reference to older siblings variously situated along the way.

These patterns were expressed, and reinforced, in the values, beliefs, laws, and customs of the culture at large. Social and ceremonial activities, no less than work, ignored distinctions of age. Church-going, family visiting, even education, regularly mixed children and adults together. Legal usage established no single "age of majority." Inheritance came most often at age twenty-one—but might also be arranged (within individual families) across a wide spectrum from fifteen to twenty-four. In some communities orphans could choose their own guardians if they were over twelve; in others, not until fourteen or sixteen. The age of criminal responsibility was also variously determined. Moreover, the significance of all numerical markers is unclear, since many people in this culture did not *know* how old they were—and others seem not to have cared.

These circumstances imply a notably fluid view of the maturational process. And when one looks to literary and philosophical materials, the same impression holds. When authors considered the life course, they did so in terms of a four-stage model. Childhood was customarily understood as lasting from birth to seven or eight years. "Youth" ended at about thirty; "old age" began at sixty. ("Middle age" was rarely even mentioned—apparently because the years between thirty and sixty seemed simply the full realization of personhood—rather than another life-stage. In a sense childhood and youth, on the one side, and old age, on the other, were construed as deviations from the midlife *norm*.)[12]

It is, of course, "youth" which primarily concerns us here; but the picture of youth which emerges from the literature is short on specifics—and flat in tone. By and large, youth was seen as a long period of gradual preparation for adult responsibilities, with few sharp twists and turns along the way. The only form of personal crisis particularly associated with youth was religious conversion. Occasionally, conversion came around the transition from childhood to youth, and more commonly it came near the end of youth. But these connections were very loosely established, and the *fact* of conversion was itself hard to pin down.[13] Puberty was not much noticed in the literature on youth; and its connotations were largely those of gradual accession to power and effectiveness. Nowhere was there any clear sense of youth as an "awkward age," with its own distinctive confusions and vulnerabilities.

To be sure, there was awkwardness (and confusion and vulnerability) in individual cases; experience rarely conforms *in toto* to cultural expectation, and did not do so here. Court records show certain specific young people in a different, more problematic light. Witch-trials, for example, might cast teen-age girls in the role of "afflicted" accusers (notoriously so at Salem); and careful study of their behavior suggests a measure of innerlife turmoil. Sometimes, too, there were youthful defendants—even groups brought into court for "nightwalking" or "company-keeping" or acts of petty vandalism.[14] Such episodes offer a hint of the "youth culture" pattern of our own day. But no more than a hint. Afflicted girls in witchcraft cases, and young people who occasionally affronted community law and morals, were exceptions that proved the general rule. And in early America the rule was relatively steady, relatively untroubled progress through youth toward adulthood.

II

After 1800 the time of life previously characterized as "youth" became increasingly disjunctive and problematic. Social circumstances, and psychological ones as well, combined to surround the passage from childhood to adult status with new elements of stress. The "adolescent" world of our own time was still a considerable distance off; but it was more and more difficult to sustain the old attitude of equanimity toward youthful experience.

One trend of central importance involved the movement of young persons—especially boys in their teens—in and out of their parental homes. Autobiographical documents from the period typically describe a sequence of comings and goings, most often on a seasonal basis. Boys of all ages were expected to be at home during the summers, in order to help their families with the heavy work of farming. But in winter and spring they might well go off to look for work elsewhere, in activities such as lumbering, or fishing, or clerking in one of the commercial towns that were sprouting so rapidly across the land. Sometimes, too, they would go to school for a season—though education was for most a very irregular process.[15] The result of all this was a new and complex web of youthful experience—in part autonomous and self-contained, in part still bound by taut cords of blood and community. One scholar has recently tried to summarize this trend by creating the term "semi-dependence."[16] (This is, of course, a half-full, half-empty proposition; thus a second scholar refers to roughly the same thing as "semi-autonomy.")[17] From the standpoint of the young person himself the change must have been quite unsettling.

For long periods he was virtually on his own, obliged simply to send back money to his family. But the old norms of filial subordination still applied: the young should in all things defer to their elders, and especially to their parents. The substance of these norms can be illustrated with a single quotation, taken from a letter written in the 1840s by a father in Maine to his twenty-five-year-old son in Boston:

> I am afraid that you do not have exercise enough to keep you in good health and spirits; and to remedy that difficulty as much as possible I should like to have you retire to rest by 9 and rise certainly by 5 in the morning, and take a long walk in the morning air at least one hour every morning when it is suitable weather. Don't neglect this, for I think it is most important.[18]

Such instructions might have seemed quite plausible in a different cultural setting (an earlier one) even for twenty-five-old sons. But in this case the young man had been living on his own for a good long time.

In fact, the first half of the nineteenth century can be seen as a time of general confusion—or at least of reshuffling—in social expectations of age-appropriate behavior. In some respects there were new elements of youthful precocity. The average age of religious conversion, for example, seems about now to have begun a steep decline. In the excitement of recurrent revivals there was pressure on the young—especially on boys and girls in their teens—to declare a personal experience of "salvation."[19] Politics was another form of excitement in the life of the new nation. By the 1820s and '30s political debate and struggle had become a major preoccupation for Americans everywhere, including young Americans. Especially at the local level, politics drew children in: often as spectators, and at least occasionally (albeit marginally) as participants.[20] Moreover, children continued to be exposed to the *emotional* concerns of their elders. Autobiographical accounts record their presence, for example, at various deathbed scenes: here is a brief, representative case. (The writer is recalling a time when, at age four, he was taken into a vault to see where his recently deceased brother and other family-members were entombed.)

> Then my kind uncle opened one of the coffins, and let me see how decayed the body had become, and told me that my

brother Edward's body was going to decay in like manner, and
at last become like the dust of the earth.[21]

In a setting where four-year-olds were encouraged to examine
decaying corpses, while twenty-five-year-olds were advised by their
fathers as to the minute particulars of their daily experience, age-
norms of all kinds were rather loosely maintained. The pattern—
or, rather, the absence of pattern—can also be seen in the records of
school attendance. In virtually all the schools of the period students of widely differing ages were mixed up together, often in a
single classroom. For example: a little "academy" in Massachusetts in 1831 reported a total enrollment of twenty pupils, and gave
their ages as follows: one student each of twelve, thirteen, fourteen,
and fifteen years old; two of sixteen; three of seventeen; one each of
eighteen and nineteen; two each of twenty and twenty-one; one of
twenty-three; and one of twenty-five.[22] There is no record of the
teacher's age in this case, but given what is known of similar
schools it could have been as low as twenty. Under such conditions
the regular maintainance of order and authority might well become problematic.

In fact, it was in the field of education that the tensions and
conflicts around issues of authority over youth revealed themselves
most fully. On the one hand, the schools and colleges of the period
displayed a minute concern with rules and regulations. Most colleges published elaborate manuals of conduct which students were
expected to learn by heart. These writings covered a broad range of
topics: study hours, dress, forms and places of recreation, and most
of all, attitudes of deference to authority. (At Yale College, for
example, the rules said that students must not wear hats when
standing within ten yards of tutors and sixteen yards of professors.)
And everywhere there was a complex scale of punishments to back
the system up.[23]

Yet it was one thing to declare the rules, and even to punish
infractions of them; it was quite another to secure from the students underlying attitudes of compliance. And the record of student *behavior* in these colleges conveys a very different impression.
The early nineteenth century was an age of student rebellion at
least equal to that of the 1960s and '70s.[24] Fistfights between
students and professors, riots over bad food, not to mention frequent combat (e.g. duels) among the students themselves: such
troubles became a commonplace of college life. As far back as the
1760s Yale had been wracked by chronic, almost continuous rebel-

lion, which eventually forced the college president to resign. At Brown University, in the 1820s, students took to stoning the president's house virtually every night. At the University of Virginia, in the 1830s, there was a particularly long and violent disturbance, culminating in the murder by students of a senior professor. The list could easily be lengthened, but the basic point is immediately clear. The strenuous disposition to force compliance (on the one side) and the urge to rebel (on the other) suggest that *all* authority relations had been thrown in question. Circumstances had reduced the capacity of the older generation to provide youth with leadership, with guidance, with assistance of the traditional kinds. The young, while in theory still expected to follow their elders, were in practice faced with new and difficult choices—to be made largely on their own.

This factor of *choice* was now emerging as key to a broad range of youthful experience: choice of occupation, choice of residence, choice of values, of friends, of sweethearts (and ultimately of spouse). Questions which, in traditional communities, had been more or less decided *for* young people were increasingly matters of individual resolution. And, where the alternatives were tightly juxtaposed, such choices and questions could be painful indeed. The predicament of a boy named Smith, age fifteen, during a series of intense religious revivals in upstate New York, was not uncommon:

> The whole country seemed affected by religious commotion, and great multitudes united themselves to the different religious parties, which created no small stir among the people— some crying "Lo, here!" and others "Lo, there!" Some were contending for the Methodist faith, some for the Presbyterian, and some for the Baptist.[25]

The boy who recalled this experience later on felt at the time "great uneasiness" and confusion, and passed many hours in the woods near his home, praying for guidance. But if the predicament was typical, its resolution in this case was most unusual. Eventually the boy received his answer: an angel appeared, instructing him not to join *any* of the competing "parties"—but instead to found a newer, and truer, religion of his own. And thus was born that quintessentially American church, known to later generations as the Mormons, with young Joseph Smith in the role of founding prophet.

By midcentury it was clear that the numerous pressures bearing in on the lives of youth had a specific locus—the new, and menacing, environment of the city. Here the choices, the temptations, the perplexities all converged. Urban youth was increasingly characterized—by the *not* young—as a social problem; a famous book about children in the 1850s and 60s was entitled, significantly, *The Dangerous Classes of New York*.[26] But not all the young people of the city were dangerous; many were themselves in danger. In direct response there flowered a whole new genre of sermons, lectures, published books and essays—variously dispersing "advice to youth." These materials, whose considerable remnant now lies gathering dust in old libraries across the country, epitomize both the problem presented to (and by) midcentury youth and the culturally preferred strategy of response. The problem, in a nutshell, was how to choose wisely and well from among the infinite possibilities thrown up by urban life; the response was moral integrity and, above all, what the advisers called "decision of character."[27] Uncertainty and vacillation must give way to willed decision: here was the central message. And if it now sounds a little thin, it must then have struck some resonant chords—for it was widely, indeed endlessly, repeated.

To be sure, the message was not directed in equal measure to *all* of American youth. More and more, as time passed, gender created boundaries of difference. During the opening decades of the nineteenth century girls in their late teens and early twenties had known their own forms of autonomy. Foreign visitors to the United States often commented on their freedom of movement (in short, the absence of chaperonage and other elements of constraint that characterized the experience of their counterparts in Europe).[28] But after about 1850 the force of gender stereotypes steadily increased. Girls approaching womanhood were seen through a haze of romanticization: to them fell the role of "junior angel in the house," from them was expected an example of gentleness, of "purity," of moral innocence. These images particularly implied suppression of self and suppression of sexuality—key elements, both, in adolescent development. And they were easily translated into socialization practice. Girls must be trained, above all, for service and self-sacrifice: sewing, cooking, and other aspects of "housewifery" made their proper curriculum.[29] Under this regime puberty itself was spotlighted as never before. For prior to puberty boys and girls shared much of everyday experience, while afterward girls entered fully into a special "female world."[30] Teen-age girls adopted the dress and manner of adult women, and they began a series of intense same-sex friendships that would color all their subsequent experience.[31]

And yet, to some extent, the earlier "freedom" accorded girls of this age survived—and took new forms. Those in the middle class (and below) were increasingly drawn out of their parental households into income-producing work. Domestic service was the most obvious possibility here (and probably the safest); and by the era of the Civil War, factory-work, school-teaching, and nursing had begun to attract large numbers of young females. Typically, such work-experience would not be of long duration; marriage would move a woman back to her appropriate domestic "sphere." Yet this sequence served to set off a period between childhood and womanhood as distinctive unto itself.[32]

In fact, for female youth, ideals and experience, norms and reality, became most painfully incongruent. Over and against the extra-familial world of work stood an elaborate edifice of conventional wisdom on biology, psychology, and social role. Organized medicine singled out the menarche as a moment of special risk and danger for girls. Then, and in the years that immediately followed, their energies would be depleted, their mental capacities dimmed, their spirits darkened. They should, therefore, in all things *be careful*—avoiding physical and mental exertion as much as possible, avoiding indeed every form of undue "excitement." A contrary pattern might doom them to lifelong illness or insanity, not to mention the likelihood of failure in woman's key role as wife-and-mother.[33]

These dark forebodings may, in many cases, have served as something of a self-fulfilling prophecy. For post-pubescent girls presented (in the second half of the nineteenth century) a bewildering range of medical problems: "vapours" and "debilities"; anorexias; neurasthenias; and, most important of all, the anemias known to physicians of the time as "chlorosis." In such clinically certified conditions—specific, as they were, to one age-group and one historical epoch—a "patient" could find refuge from the conflicting pressures of cultural stereotype, innerlife development, and real-world experience.[34]

III

The common denominator in the experience of nineteenth-century youth—common to boys *and* girls—was dissonance. But the end of the century brought a counter-trend, a measure of codification and confinement. This indeed was the period of G. Stanley Hall's most important work and the "dawning" of the modern concept of adolescence.

Many sectors of American life in the late nineteenth century were characterized by what one scholar has called a "search for order." And the same impulse appears in the experience of young people vis-à-vis their elders. Individually and community-wide, efforts were made to channel, to organize, and thus to constrain the jarring influences of a generation or two before. Consider, for example, the matter of education. It was then, and only then, that primary schools and high schools emerged as distinct institutional units. The result was an end to the old pattern of mixing ages in the schools; from henceforth the young would be literally *graded* as they passed through their student years. But the efforts of schools in this connection were complemented by trends internal to families. The earlier forms of "semi-dependence"—whereby youth moved back and forth between family settings and the wider community—gradually disappeared. From henceforth youth was to be a time of fuller and fuller—hence longer and longer—dependency.[35] Demographic change also played in here. The birth rate had been falling throughout the century, so that average family size was reduced by almost half. Concomitantly, and not coincidentally, there developed a new style of careful, highly self-conscious parenting. (Another historian finds in this period a whole ethos of "intensive family life.")[36] Mothers, in particular, were expected to devote themselves to individualized nurture of their children. The larger impulse which underlay all such activity, whether within or outside the home, was to create systematically planned environments for the young. Youth might then progress in a more orderly fashion toward the great goal of adulthood, bypassing insofar as possible the perils and pitfalls along the way.

In a sense the outcome of these trends was a broad-gauge standardization of youthful experience. From henceforth development was subject to a kind of social processing, with schools as the key institutional device, but with other institutions also playing their part. There were, for example, new organizations devoted to the improvement of both morals and practical skills in the young: the Boy Scouts and Campfire Girls, the Y.M.C.A. and Y.W.C.A., the 4-H clubs, the fraternities and sororities, and so on.[37] Adolescent fashions—in dress, in recreation, in speech and social mannerisms—reflected the same trend; indeed it is to this period (around the turn of the century) that we can trace the appearance of a true "youth subculture" with features familiar to all of us today.

The effort to standardize the lives of young people was ambivalent in motive and complex in result. On the one hand, there was some intention to suppress precisely those elements of experience which seemed most "youthful"—the adventuresomeness, the opportunity to sample a broad range of behavior-styles and beliefs and values. On the other hand, there was implicit recognition that adolescence was a *bona fide* life-stage with its own particular requirements. And insofar as the standardizing tendency brought young people together in settings specifically contrived for them (and them alone), it greatly accentuated peer-group consciousness. To be an adolescent was to share with others of similar age not only a developmental position but also a social status. Peer solidarity became a force in its own right, and young people began to claim certain things *because* they were young. Consider, in this connection, a schoolteacher's wry complaint to Professor Hall about the effect of his theories on her students: whenever she reproached them for misbehavior in class, they would reply, "Oh, but teacher, we're passing through adolescence!"[38]

Moreover, there were powerful trends in the culture at large that may well have served to intensify adolescent self-consciousness. For one thing, there was immigration on a completely unprecedented scale—a million or so new arrivals each year, some in family-groups, the rest soon to start families in their adopted home. As these families grew and changed, they experienced a particularly raw form of the generation-gap, with parents still anchored to some of the ways of the Old World, and their children freshly attuned to the life of the New. There was also a second kind of immigration in turn-of-the-century America, bringing millions of young people, born and partly bred in rural or small-town settings, into the burgeoning life of the city. This, too, was a fertile source of cross-generational tension, with evident implications for adolescence.

But these are large historical speculations, hard to pin down within the compass of a single, short discussion. And so it seems best to call a halt just here, or at least to direct attention to the residual *psychological* aspects of the material. Adolescence—we have seen—became during the nineteenth century an increasingly sharp and problematic life-stage, for an increasingly large number of young Americans. The question, framed now in psychological terms, remains *why*. Various possibilities suggest themselves. We might look at the matter from the vantage point of "classical" psychoanalytic theory—which posits, as a central feature of adolescent experience, the revival and reworking of early "oedipal" themes. In fact, many elements of nineteenth-century social history had converged to render such themes conspicuous throughout the culture. Smaller and smaller households, increased sex-typing *within* households, the development of particularly intensive styles of child-nurture: all this had served to create a kind of "oedipal family" in middle-class America. And so: if oedipal issues had become bigger and tougher *to begin with,* then adolescence might well have begun to exhibit heightened levels of what contemporaries called "storm and stress."[39]

One could also point this discussion toward psychoanalytic "ego" psychology: for example, toward Erik Erikson's well-known work on "identity."[40] We have seen how life-choices were multiplied in nineteenth-century America, and how, too, the experience of one generation was increasingly detached from the prospects of the next. This is to say, in other words, that many adolescents found themselves caught up in "identity diffusion" of a sort hardly conceivable during earlier historical periods. The process of deciding who one was, what one wanted to be, and how one's particular choices would intersect with the social order: here was a labyrinth of personal—and social—perplexities.

IV

What of adolescence in more recent times—and in our own time? The question has, of course, been studied, and analyzed, and worried over, by a veritable army of contemporary authorities. The resultant literature would fill a library all of its own. Yet an historian's "long view" finds more continuity than change in twentieth century adolescent experience, more elaboration of established trends than creation of new ones. Personal "crisis," on the one hand, and social "channeling," on the other: these have remained dominant themes from G. Stanley Hall's time almost to our own.

But perhaps not all the way to our own. There are new patterns and portents on every side—but so new that it is impossible to present them in a scholarly way. Thus an historian must beg his reader's indulgence for a radical shift of viewpoint. What follows is simply his own impression—subjective, suggestive, personal, designed more to provoke than to prove. . . .

The historian has lived most of his adult life on a small street in a New England suburban community. And on this street he has watched an entire cohort of children move from roughly the beginnings to past the end of their school-age years. He knew them first as toddlers and kindergartners who filled the neighborhood with their noise and games and general high spirits. They seemed open, curious, imaginative, and, above all, enviably free in their emotional experience. A day, or even an hour, might put the full range on view: joy, anger, fear, excitement, boredom, shame, surprise. Then time passed, and the children became a few years older—ten, eleven, twelve—"pre-adolescent" in the jargon of the learned professions. Their lives and their play, at least as manifested in the neighborhood, assumed a more organized and purposeful aspect: they moved from backyard sandboxes and swing sets into the middle of the street, with games like stickball and soccer and asphalt hockey. But their style remained in all ways open and emotionally expressive.

Then they became teen-agers, and everything changed. First of all, they completely withdrew from the life of the street: one scarcely knew they were present anymore, still a part of their various family households. Moreover, a new set of goals and values seemed to have taken hold of them, and their daily activities were radically transformed. Of course, they still attended school—high school, by this time—but after hours, almost without exception, they went off to some form of income-producing work. They could be found bagging groceries in the nearby supermarket, or pumping gas on weekends at the filling-station, or painting and doing roof-work with local contractors. From all this they brought home *money*, which was soon invested in cars, clothes, stereos, and the like; occasional conversation suggests that they are nearly as much involved in monthly payments and credit arrangements as their parents. Indeed, the older ones, having finished high school, are moving into full-time employement as secretaries, receptionists, clerks, and factory-workers. And, too, they are getting married and starting to have children of their own—at nineteen or twenty years of age. There is, finally, the question of their emotional style: here the change has been especially remarkable and disturbing. In a word, the freedom, the openness, the imaginative and expressive range that was so prominent in their early years seem virtually gone; there is something flat, stolid, altogether routinized, about them nowadays. Far from experiencing adolescence as a time of varied opportunity, of experimentation, of flux, of "storm and stress" (on the old model of G. Stanley Hall), they have narrowed their sights in a very marked way, and indeed have closed off important options and possibilities. No doubt it would be too simple to say that they have skipped adolescence entirely, but they do seem to have bypassed (what the historian remembers as) the typical experience of their counterparts a generation ago.

These young people belong to what sociologists would call the "lower middle class," and perhaps the differences sketched above are less a function of change over time—the 1950s *versus* the '70s and '80s—than of differential *class* experience. No doubt class position has always influenced the forms the life course takes in individual cases; and this should be as much true of adolescence as of any other stage. Still, it does not respond very fully to the evidence specifically at hand. For the historian feels a similar tendency in his classroom—in his *students*, also adolescents, but largely representative of the upper-middle class. In fact, the "conservative" temper of the present college cohort has been widely remarked on. They, like their non-college peers, seem uninterested in adventure and experiment. Their goals, too, run to security and material comfort, and their day-by-day experience is a straightforward program of pre-professional advancement.

This is, of course, a drastic over-simplification, but it helps to plot the broad lines of historical change. And the change that is finally in question here is nothing less than a change in the shape of the life course. Adolescence may no longer be as stressful, as variable, as altogether *salient*, as was the case mere decades ago. There is at least some support for this proposition in the recent literature of social science—in the work, for example, of the psychologist David Elkind, the psychoanalyst Daniel Offer, the sociologist James Coleman, and the educator Edgar Z. Friedenberg (whose prescient and passionate book *The Vanishing Adolescent* was published as far back as 1958).[41]

Historians, in particular, might take this as a caution. We are accustomed to believe that the life course has become more various and differentiated with the passing of the centuries—that new "stages," new age-specific forms of experience have been added, under the transforming influence of historical circumstance. This appears to be another way of saying that life has been growing richer, fuller, more interesting, and that development itself has been developing in a favorable direction. Here is another aspect of what might be called the "whig interpretation" of family history.

Yet our own time offers us some different, if not contrary, evidence. Life-stages may come and go, or at least sharpen and fade, even within a relatively short span of time. In family history—as in other historical sub-fields—there are no straight lines.

NOTES

1. See John and Virginia Demos, "Adolescence in Historical Perspective," in *Journal of Marriage and the Family*, XXXI (1969), 632–38; Joseph Kett, "Adolescence and Youth in Nineteenth-Century America," in *Journal of Interdisciplinary History*, II (1971), 283–298; John Demos, *A Little Commonwealth: Family Life in Plymouth Colony* (New York, 1970), ch. 10; Philippe Ariès, *Centuries of Childhood: A Social History of Family Life,* trans. Robert Baldick (New York, 1962), ch. 1; and John R. Gillis, *Youth and History: Tradition and Change in European Age Relations 1770–Present* (New York, 1974).

2. On Hall's life and career the definitive work is Dorothy Ross, *G. Stanley Hall: The Prophet as Psychologist* (Chicago, 1972).

3. G. Stanley Hall, *Adolescence* (New York: D. Appleton & Co., 1904)

4. The full title (with sub-title) is *Coming of Age in Samoa: A Psychological Study of Primitive Youth for Western Civilization* (New York, 1928).

5. See, for example, Kingsley Davis, "The Sociology of Parent-Youth Conflict," in *American Sociological Review*, V (1938), 523–35.

6. This "revisionist" viewpoint on the history of adolescence can be sampled in the following works: Lawrence Stone, *The Family, Sex, and Marriage in England 1500–1800* (New York, 1977), 375–77; Natalie Zemon Davis, *Society and Culture in Early Modern France* (Stanford, Calif. 1975), esp. ch. 4; Stephen R. Smith, "Religion and the Conception of Youth in Seventeenth-Century England," in *History of Childhood Quarterly*, II (1975), 493–516; and Vivian Fox, "Is Adolescence a Phenomenon of Modern Times?," in *Journal of Psychohistory*, V (1977), 271–90. For specific applications to early American history, see: Ross Beales, "In Search of the Historical Child: Miniature Adulthood and Youth in Colonial New England," in *American Quarterly*, XXVII (1975), 379–98; N. Ray Hiner, "Adolescence in Eighteenth-Century America," in *History of Childhood Quarterly*, III (1975), 253–80; and Roger Thompson, "Adolescent Culture in Colonial Massachusetts," *Journal of Family History*, IX (1984), 127–44.

7. The most balanced book-length treatment of this subject is certainly Joseph Kett, *Rites of Passage: Adolescence in America, 1790 to the Present* (New York, 1977).

8. See Hall, *Adolescence*, vol. I, 26–28, and J. M. Tanner, "Sequence, Tempo, and Individual Variation in the Growth and Development of Boys and Girls Aged Twelve to Seventeen," in *Daedalus* (Fall, 1971), 907–30.

9. See *ibid.*, and Peter Laslett, "Age at Menarche in Europe since the Eighteenth Century," in *Journal of Interdisciplinary History*, II (1971), 221–36.

10. For a fuller discussion of these patterns in early American childhood, see Demos. *A Little Commonwealth*, 57–58, 131–44; and Michael Zuckerman, *Peaceable Kingdoms: New England Towns in the Eighteenth Century* (New York, 1970), 73ff.

11. The issues summarized in this paragraph are themselves the nub of a scholarly controversy. For a view that makes much more of child/adult differences in early America, see Beales, "In Search of the Historical Child."

12. See Chapter Six, below.

13. On the timing of religious conversion in early America, see Robert G. Pope, *The Half-Way Covenant* (Princeton, 1969); Philip J. Greven, Jr., "Youth, Maturity, and Religious Conversion: A Note on the Ages of Converts in Andover,

Massachusetts, 1711–1746," in *Essex Institute Historical Collections,* CVIII (1972), 119–34; and other studies noticed in Patricia J. Tracy, *Jonathan Edwards: Religion and Society in Eighteenth-Century Northampton* (New York, 1980), 225-26 (fn. 37).

14. For details of such episodes, see Thompson, "Adolescent Culture in Colonial Massachusetts," Thompson's view of their *significance* is, however, considerably at variance with the argument presented here.

15. For specific evidence on these points, see Kett, *Rites of Passage* ch. 1. Comparable material for nineteenth-century Canada is presented in Michael Katz, *The People of Hamilton, Canada West* (Cambridge, Mass., 1978), ch. 5.

16. Kett, *Rites of Passage,* 29.

17. Katz, *The People of Hamilton, Canada West,* 256.

18. Quoted in Charles W. Moore, *Timing a Century: History of the Waltham Watch Company* (Cambridge, Mass., 1945), 5.

19. This topic is discussed at length in Kett, *Rates of Passage,* ch. 3.

20. "Genre painters" of the period often portrayed scenes of political activity—with children specifically included. See, for example, the work of George Caleb Bingham, and the discussion in John Demos, "George Caleb Bingham: The Artist as Social Historian," in *American Quarterly,* XVII (1965), 218–28.

21. As recalled by Samuel J. May, and quoted in Lewis O. Saum, "Death in the Popular Mind of Pre-Civil War America," *American Quarterly,* XXVI (1974), 477–96.

22. See Kett, *Rites of Passage,* 20.

23. *Ibid.,* 51ff.

24. On student disorders, see David F. Allmendinger, Jr., *Paupers and Scholars: The Transformation of Student Life in Nineteenth-Century New England* (New York, 1975), ch. 7.

25. Quoted in Kett, *Rites of Passage,* 69.

26. The author was a prominent Victorian reformer and philanthropist, Charles Loring Brace. The book's full title was *The Dangerous Classes of New York and Twenty Year's Work Among Them* (New York, 1872).

27. This literature is well summarized in Kett, *Rites of Passage,* 102-4, 161–62.

28. Alexis de Tocqueville, for example, commented that "nowhere are young women surrendered so early or so completely to their own guidance" as in the United States. See his *Democracy in America,* the Henry Reeve Text, Phillips Bradley, ed. (New York, 1956), II, 209.

29. The summary presented here is based principally on Deborah Gorham, *The Victorian Girl and the Feminine Ideal* (Bloomington, Ind., 1982); Carol Dyhouse, *Girls Growing Up in Late Victorian and Edwardian England;* and Patricia Branca, *Silent Sisterhood: Middle-Class Women in the Victorian Home* (London, 1975). These studies are directed, for the most part, to nineteenth-century England. Unfortunately, there is no comparable treatment of the experience of American girls, but their situation was at least *similar.* Bits and pieces of the American story can be found in: Carroll Smith-Rosenberg, "The Female World of Love and Ritual: Relations Between Women in Nineteenth-Century America," in *Signs,* I (1975), 1–29; Smith-Rosenberg, "Puberty to Menopause: The Cycle of Femininity in Nineteenth-Century America," in *Clio's Consciousness Raised,* Mary Hartman and Lois W. Banner, eds. (New York, 1974), 23–37;

Nancy F. Cott, *The Bonds Of Womanhood: "Woman's Sphere" in New England, 1780–1835* (New Haven, Conn., 1977); Carl Degler, *At Odds: Women and the Family in America From the Revolution to the Present* (New York, 1980); Mary P. Ryan, *Cradle of the Middle Class: The Family in Oneida County, New York, 1790–1865* (Cambridge, Eng., 1981); and Phillida Bunkle, "Sentimental Womanhood and Domestic Education," in *History of Education Quarterly*, XIV (1974), 18–35. I should like also to acknowledge my debt to Professor Joan Jacobs Brumberg (Cornell University) for valuable suggestions about the history of female adolescence.

30. See the discussion in Gorham, *The Victorian Girl*, 85–87. One medical authority described the differential impact of puberty in the following way: ". . . that which makes men more bold will generally awaken greater timidity in women. Puberty which gives man the knowledge of greater power, gives to woman the conviction of her dependence" (Edward John Tilt, *The Elements of Health, and Principles of Female Hygiene* [London, 1852], 173; quoted in Gorham, *The Victorian Girl*, 86).

31. *Ibid.*, ch. 6. On the matter of "homosocial" friendship among nineteenth-century women, see also Smith-Rosenberg, "The Female World of Love and Ritual."

32. For an excellent summary of women's work in the nineteenth-century, see Degler, *At Odds*, ch. 15. See also: Gorham, *The Victorian Girl*, 28–31; Dyhouse, *Girls Growing Up in Late Victorian and Edwardian England*, 82–84; Branca, *Silent Sisterhood*, 54–57; Kett, *Rites of Passage*, 95–96; Katz, *The People of Hamilton, Canada West*, 270–72.

33. See: Smith-Rosenberg, "From Puberty to Menopause;" Elaine and English Showalter, "Victorian Women and Menstruation," in *Suffer and Be Still: Women in the Victorian Age*, Martha Vicinus, ed. (Bloomington, Ind., 1972), 38–44; Gorham, *The Victorian Girl*, ch. 5: Dyhouse, *Girls Growing Up in Late Victorian and Edwardian England*, 21–22, 131–38. G. Stanley Hall himself asserted a radical distinction between male and female adolescence. A useful summary of Hall's views on this point can be found in Dyhouse, 122–28.

34. On the medical problems of adolescent girls in this era, and especially on the syndrome called "chlorosis," see Joan Jacobs Brumberg, "Chlorotic Girls, 1870–1920: A Historical Perspective on Female Adolescence," in *Child Development*, LIII (1982), 1468–77. See also Gorham, *The Victorian Girl*, 88–97.

35. These trends are discussed at length in Kett, *Rites of Passage*, chs. 5–6. See also Katz, *The People of Hamilton, Canada West*, ch. 5. (esp. pp. 273–78, 290–92).

36. Stephen Kern, "Explosive Intimacy: Psychodynamics of the Victorian Family," in *History of Childhood Quarterly*, I (1974), 437–60.

37. On boys' organizations, see Kett, *Rites of Passage*, ch. 8; on girls, see Dyhouse, *Girls Growing Up in Late Victorian and Edwardian England*, 106–114.

38. Quoted in Walter D. Hyde, *The Quest for the Best* (New York, 1913), 1.

39. On "oedipal" themes in turn-of-the-century American family-life, see John Demos, "Oedipus and America: Historical Perspectives on the Reception of Psychoanalysis in the United States," in *The Annual of Psychoanalysis*, VI (1978), 23–39.

40. See, for example, Erik H. Erikson, *Identity: Youth and Crisis* (New York, 1968).

41. David Elkind, *All Grown Up and No Place to Go: Teen-agers in Crisis* (New York, 1984); Daniel Offer, Eric Ostrov, and Kenneth I. Howard, *The Adolescent: A Psychological Self-Portrait* (New York, 1981); James S. Coleman, *The Adolescent Society* (New York, 1961); Edgar Z. Friedenberg, *The Vanishing Adolescent* (Boston, 1959). Kett makes a somewhat similar argument in the concluding chapter of his *Rites of Passage* (see esp. 266ff).

CHAPTER VI

Towards a History of Mid-life: Preliminary Notes and Reflections

By the late 1970s historians had taken up "life course" studies in a remarkably vigorous way. Infancy, childhood, adolescence, old age, not to mention the special moments of birth and of death: these matters furnished more and more grist for a larger and larger scholarly mill. Curiously, however, one part of the life course remained as yet untouched: the part we call mid-life or, simply, middle age. Perhaps, I speculated, this had something to do with where most historians are in their own life course?

But then again, perhaps not; my own instinct ran very much the other way. I had just celebrated (?) my fortieth birthday when the idea for the next essay occurred to me. Here—in contrast to all other parts of the present volume—there was no outside invitation or assignment; the inducements were wholly internal. I wanted, I welcomed, the chance to study the history of middle age.

In truth, there are other, quite impersonal, reasons why historians might avoid this particular subject. For one thing, it does not easily define itself in conceptual terms; for another, it does not declare its presence in the extant historical record. With other lifestages, a presence of some sort can usually be established—with childhood, certainly, with youth (if not adolescence in our sense), with old age. But middle age is mostly a non-presence; is it therefore a non-subject? The answer must be mixed: yes and no—or, rather, no and yes. No, in that human beings of every society, every epoch, must perforce have passed through a time of life equivalent (at least in chronological terms) to our middle years. And, yes, in that none of them before our own era made very much of the fact. It is really only in the last two or three decades that midlife has been identified, and analyzed, as a distinct developmental phase. Of course, such identification cum analysis cannot be directly equated with changed experience; but neither are the two things unrelated. Something is changing: it remains to discover exactly what.

What follows is but a step in that direction. Lacking any other historical studies of mid-life—lacking, too, much overt assistance from the historical actors themselves—I have been forced to read back from present-day experience, and out from present-day theory. Eventually, it will be necessary to build this history up, piece by painstaking piece, through the study of individual lives. But I have managed to put only a few such pieces in place here. I entreat other middle-aged historians to do more.

In middle age, writes one recent commentator (with her tongue only partially in her cheek),

> the hormone production levels are dropping, the head is balding, the sexual vigor is diminishing, the stress is unending, the children are leaving, the parents are dying, the job horizons are narrowing, the friends are having their first heart attacks . . . The past floats by in a fog of hopes not realized, opportunities not grasped, women not bedded, potentials not fulfilled, and the future is a confrontation with one's own mortality.[1]

This bleak sketch is meant, evidently, to capture *male* middle age. To women it would apply much less—or not at all. But for middle-aged men, especially those of the middle class in North America in the late twentieth century, it does seem to hit the mark.

Indeed, we have a modish, late twentieth-century term for the entire package: the oft-mentioned "mid-life crisis." And we have a veritable industry of scientific research on the subject—which, in turn, has spawned a huge literature of a more popular sort.[2] Beneath so much sound and fury, behind so much "smoke," burn the proverbial fires of experience; clearly, it would be wrong to dismiss "mid-life crisis" as a whim or fad of no serious consequence. The effects, not to say the phenomenon itself, can be gauged in rather precise quantitative terms: in rates of divorce, of career-change, of psychiatric office-visits, of alcoholism and suicide. The qualities are harder to measure, but they have a cultural resonance that seems in itself compelling.

But is the resonance specific to *our* culture here and now? Or is there something transcultural—which is to say invariant and unavoidable—about mid-life experience? A leading theorist of the life cycle "energetically" offers the following hypothesis: The same "sequence of eras and periods [in middle age] exists in all societies, throughout the human species, at the present stage in human evolution. The eras and periods are grounded in the nature of man as a biological, psychological, and social organism."[3] A historian, however, feels an almost instinctive skepticism in the face of such sweeping affirmations. He may agree that certain biological features of the aging process do operate transculturally—for example, menopause in women, and a diminution of physical resiliency in both men and women. He may even agree that some psychological elements are everywhere present—for instance, growing intimations of personal mortality. But he is *not* likely to agree that "social" factors also weigh on the side of universality. Indeed, all his professional knowledge declares the opposite. Society as such seems to him infinitely variable—seems, moreover, to condition biology and psychology, or at least to control their meaning. An event like menopause, and a process like physical weakening, can be experienced in enormously different ways, depending on the values, the customs, the habits of thought, the contingent material, demographic, and ecological circumstances of the people most directly involved. What is a "crisis" in one setting may wear a very different aspect in another.

The present discussion explores the interaction between biology, psychology, and society—between the allegedly universal and the apparently variable elements in human experience—with explicit reference to middle age. And it does so by way of a single then-*versus*-now comparison. "Then" is, in all that follows, the "pre-modern" culture of seventeenth-century New England; "now" is our time and culture. As with all historical comparison, the aim is to illuminate both settings, both patterns of experience. In learning more about the past we also manage—if all goes well— to learn more about ourselves.

I

Immediately, the record of early New England presents us with a *conundrum*. When we look in that record for the markings of our subject—for some explicit recognition of middle age—we seem at first to come up empty. Indeed, the very term is missing from the discourse of these, our remote ancestors. Other "developmental" terms they had, and used—but not this one. Thus, for example, when the Reverend Cotton Mather of Boston decided to publish a series of sermons on the personal side of religious experience, he produced a volume entitled *Addresses to Old Men, and Young Men, and Little Children.*[4] This, it appears, expressed the prevailing view of the life course in Mather's time. Childhood, as he and his cultural peers defined it, lasted until about the age of six or seven. "Youth," which was seen as a fairly straightforward period of preparation for adult responsibilities, carried forward to perhaps age thirty. And old age began at sixty. Between thirty and sixty stretched a time of life which Mather and others saw little reason to discuss, and which they scarcely bothered to name. It appears, in fact, that they did not regard this interim period as distinctive at all; instead, the middle years represented for them simply the full flowering of human capacities. Someone in his thirties, forties, or fifties was a fully developed *person*—a norm against which childhood and youth, on one side, and old age, on the other, could be measured as deviations. This entails a significantly different view of the life course from our own—a view less fully "linear," because in some sense discontinuous.*

It also reflects a different attitude toward the middle years. Our term "middle age" expresses our sense that this is a well-articulated stage of life, no less so than other stages which precede and follow it. And when our notion of "crisis" is added in, the then/now contrast becomes bald indeed. Therein lies both a problem and an opportunity. The problem, of course, is one of interpretation: how to explain such striking differences in human perception—and human experience? The opportunity is to explore the

---

*Iconographic evidence from pre-modern and early modern Europe typically represents the life course as a kind of hill, with ascending and descending slopes on either side of (what we would call) middle age. An example from eighteenth-century London shows a male/female pair at each ten-year interval from birth to age 100; the apex is set at 50. The apex figures are different from all the others—are faced frontward, are larger and more imposing, hence are clearly designed to reflect maximum power and status. (See the illustration in John R. Gillis, *Youth and History* [New York, 1974], xvi.) There is no comparable material from early New England—but this little paradigm would apply there just as well.

whole complex interaction between history and life-history, between the broad flow of time and events and the way all of us as individuals experience our own little parts of that flow.

II

It would be too much to claim that current research yields a rounded, overall theory of development in the middle years. Still, it does set in place the foundation-stones for such a theory. And these, in turn, may well serve as fixed points from which to extend and refine our view of the past.

There are five distinct elements in the social science literature that bear consideration here. Together they purport to cover much of the stress which attends middle age; briefly listed, they are as follows: (1) a heightened level of concern with death, and, more specifically, with the inevitable fact of one's own mortality; (2) a progressive loss of physical strength and vitality; (3) a decline in sexual interest and activity; (4) a confrontation with aspects of one's inner life and personality which had hitherto remained dormant, if not repressed outright; and (5) a larger process of reassessment of the life one has lived so far, with a view toward some reshuffling and reordering for the time that still lies ahead. Each of these processes, with the possible exception of the last one, has some roots in the area of biological and biochemical change—although the precise relationships, in terms of hormones, cell changes, and the like, are apparently not well understood. At the same time, each one is mediated by social factors and influences, which may serve either to intensify or to diminish the strength of a given effect, and to supply much of its experiential content.

Let us begin with the element of confrontation with death. There is abundant evidence, both from clinical work and from everyday experience, that people in middle age become deeply preoccupied with their own finitude—with the inescapable realization that their time is limited, and that this limitation is largely beyond their power to control.[5] The facts of biology and demography underlie such preoccupations, but consider how powerfully the historical setting may influence their shape and strength.

In colonial America, for example, death was a substantial presence in all areas and for all ages of people. Most families experienced the loss of one or more children to measles, smallpox, dysentery, or other ubiquitous, ill-defined "fevers." Some mothers died in childbirth. Even young men, seemingly at the height of their physical powers, occasionally succumbed to epidemic disease. The total of mortality was less than is sometimes imagined, and it was clearly less than in most parts of Europe during the same period—but was still far greater than anything we know today. Moreover, these losses were distributed throughout the life cycle of those who survived. The average person experienced them at more or less regular intervals from his earliest years. Finally, demographic reality was effectively reinforced by cultural norms and conventions. The prevalent systems of religious belief—with "Puritanism" as the most notable instance—encouraged a posture of candid and continuous attention to death. Even small children were exhorted to "remember death, think much of death, think how it will be on the death-bed" (the words are those of a seventeenth-century New England schoolmaster).[6]

Consider, by contrast, the prevalent situation in our own time and culture, where encounters with death are limited, postponed, denied, though not, of course, finally avoided. A personal case—the writer's own experience—may be illustrative. Death entered his life as a child hardly at all; he does recall the passing of a grandmother and an elderly uncle, and the sudden disappearance of one terminally ill playmate, but these were faraway events from which he was emotionally shielded by his parents. The first real losses of peers occurred when he was in college or shortly afterward; but the cause was suicide, which placed the matter in a special light. In short, he had no full and direct experience of "natural death" during all those years—not until his father died when he was thirty-one. (His father's funeral was actually the *first* funeral that he had ever attended.) Since then the picture has changed dramatically. Now, as he reaches his mid-forties, many other members of his parents' generation are gone, and at least a few in his own age group, too. Moreover, others among his peers have been shadowed by potentially mortal illness.

The particulars in any individual case are, of course, induplicable, but broadly speaking this is the pattern of many lives nowadays. Death is simply not much of a factor in the childhood years, or even in the years of early adulthood, but it asserts itself quite forcibly as we reach our thirties and forties. Hence for us in modern America the experience of death as the loss of loved ones and friends is age-specific: it bears down on us in middle age—when we are already predisposed for psychological reasons to consider our own mortality. Inward tendency is thus reinforced by an outward sequence of events. And such reinforcement was largely missing from the lives of our forebears two and three centuries ago.

A parallel contrast appears around the physical decline that also characterizes middle age. This, too, is biologically inevitable—but in our time is highlighted by cultural patterns and events. In fact, the pace of physical decline can be slowed by regular exertion and exercise, or speeded by a more sedentary style of life. The difference rather neatly divides the lives of many of us from the regime that was typical of an earlier time. The overweight businessman who looks back on his achievements as a college athlete twenty years before cannot help but notice the signs of physical depletion; not so, or at least not to the same extent, for the forty-year-old farmer whose life is spent in regular physical activity. Paradoxically, the forms of exercise which some of us choose as ways of "staying in shape"—in effect, arresting physical aging— are just those which reveal even marginal decline. To find oneself a step slower on the tennis court, or gradually slackening the pace on a daily jogging excursion: such things are painful and hard to overlook. In competitive sports the score, the stopwatch, the mile-markers precisely measure the change.

One final consideration of some importance here is the relative freedom from illness that most of us can expect in young adulthood. Even for the most inactive serious health problems can generally be warded off until at least the mid-thirties; then come the coronaries, the ulcers, the back problems, the high blood pressure, and all the rest. In pre-modern times, by comparison, health and illness seem to have been more evenly distributed through the life-span. The major threats to health—most especially, the many varieties of epidemic disease—affected children and young adults no less than people in the later stages of life. Again, *our* pattern, far more than *theirs*, accentuates the problematic factors in middle age.

Sexual decline is perhaps just a piece of overall physiologic decline; still, it is widely experienced as a distinct—and alarming—process in its own right. The facts of this matter have been made abundantly clear in numerous scientific studies from Kinsey right through Masters and Johnson: sexual activity, at least as measured in quantitative terms, traces a downward curve beginning in most persons by the mid-thirties.[7] This factor receives special attention in psychoanalytic theories of the aging process: in short, decreased levels of sexual activity signify, in the realm of the unconscious, a loss of potency, and this reactivates infantile fears of castration.[8] In some people—one should candidly say, some *men*—the result is manifestly "counter-phobic"; that is, there begins a frantic and compulsive search for new sexual adventures, which may help to ward off the underlying anxieties.

Again, inquiry turns from present to past time, but for this issue there are particularly grave problems of evidence. We have precious little evidence of the sexual behavior of men and women in earlier times, and no evidence at all of their innermost fears and fantasies. Still, it is possible to draw the outlines of a contrast, at least in speculative terms. Our culture is notoriously interested in sex; more specifically, we are interested in sexual *performance*. Perhaps it is not too much to say that we treat such performance as an important aspect of self—and of self-esteem. The people of colonial America, on the other hand, had no such feelings. Perhaps it will seem odd to think of these so-called Puritans as more natural, more relaxed, less self-conscious in their sexual attitudes than their modern-day descendants; but that is, in fact, the central finding of a good deal of recent research in sex history.[9]

There is a further point as well. The average "Puritan" man or woman continued with childbearing until, or even past, the age of forty; as a result, there were small children underfoot in many colonial households until the parents were in their fifties. Perhaps this continuing presence of young progeny helped to mute the sense of sexual decline. In our own case, by the time parents reach middle age child-*bearing* is long since over—and the children are reaching adolescence. And the awakening sexuality of a fifteen-year-old son, for instance, is sometimes experienced by a forty-year-old father as a kind of challenge—an implicit reproach which calls attention to his own declining powers.

We may turn at this point to some rather different, though not unrelated, forms of psychological aging. There is a body of recent cross-cultural research which shows markedly contrasting inner-life tendencies in men and women as they pass through middle age. Very briefly, the gist is as follows. Men undergo a steady shift, away from an active, assertive, striving orientation toward their everyday world and into something that gives more scope to their affiliative, nurturant—may one even say "feminine"?—side. For women the process seems to be precisely opposite: that is, *away* from affiliation and dependence, *toward* aggressive self-assertion. In short, in middle age there begins a process of psychological convergence of male and female—or even a swapping of positions, an outright crossover. Moreover, since the process shows up in a variety of otherwise different cultures, it lays claim to a universal developmental significance. It can be demonstrated for our own culture as well, with data gathered from clinical interviews; and it appears, in at least attenuated forms, as an aspect of ordinary experience—for instance, the sudden interest taken by many older men in pursuits such as gardening.[10]

But the larger point about our culture is this: on the whole, our values and social arrangements are unreceptive to fundamental sex-role change. If a middle-aged man experiences, at some level, a sense of incipient "feminization," he and his peers cannot rest comfortably with this; likewise for a middle-aged woman who begins to feel the opposite tendency. In other cultures the situation is sometimes very different; there one finds support, even a kind of "sponsorship," for the same inner changes. Sex-roles alter sharply in accord with age, and new but socially rewarding activities are presented to the midlife cohort. Men, for example, may move from roles as warriors and hunters to those of priests, seers, and elders, while women exchange domestic care-taking for entrepreneurial activities in the local market.[11] But is this a general "pre-modern" pattern, one that holds even for early America? Unfortunately, the New England evidence is too sparse to support an answer one way or the other; hence the matter must be left in the form of a question. However, the orientation of our own society to these processes does seem clear—and does, once again, weigh strongly on the side of stress and conflict in middle age.

The final problem-area to be mentioned here is the element of personal reappraisal that seems now so normative for middle age. There is a pervasive sense of options and possibilities passed by, of choices too irrevocably made, of limits closing in.[12] In a word, the grass begins to look much greener on the other side of a lot of fences, even as the fences themselves come to seem impassably high. Almost certainly, these preoccupations are rooted in the experience of finitude (noted above)—the feeling that only so much time is left, too little, presumably, for doing all that one might wish to have done.

Insofar as finitude is the bottom line, there must be some universal involvement here, irrespective of cultural and historical settings. Perhaps a very sensitive review of religious literature in colonial America—private journals of self-examination, sermons in response to special events—would disclose at least glimmerings of middle-aged reappraisal. However, it would not add up to a great deal; once again, the important point is the relative absence *then* of something that bulks very large in our own time. The difference, briefly summarized, is the whole dimension of *choice* in our lives—as contrasted with theirs. In fact, this is the same element that contributes so powerfully to our problematic experience of adolescence, though in mid-life the problem is not how and what to choose, but how to live with the choices already made.[13] The situation is necessarily compounded when there appear to be so many choices not made, so many alternatives missed or simply ignored. For our colonial ancestors, the middle years must have been easier—because their alternatives were fewer, their verities less open to challenge or change.

III

The main line of this discussion so far has drawn a comparison between the experience of midlife in two historical settings separated from one another by some three hundred years in time. But, of course, the terms of the comparison are somewhat skewed: we have started from the distinctive features of middle age *now*, and searched—largely in vain—for similar signs and tendencies in the lives of the settler generations. And if we have learned something by this method about what midlife in early America was *not*, we are as yet unable to say what it *was*. It is important, then, to make some effort to fill the blank spaces.

We began, a good many pages ago, by remarking that there is no sign in the source materials from early America of a sense of mid-life as a developmental phase at all. And that remains an accurate generalization overall—and a revealing one. But there is at least one exception to it which does bear considering: a long writing in verse, by the Puritan poet Anne Bradstreet, entitled "The Four Ages of Man." Bradstreet begins:

> Lo, now, four others act upon the stage,
> Childhood and youth, the manly, and old age . . .

and proceeds to characterize each of these allegedly typical "ages" in considerable detail. Even here at the outset she has a significant way of identifying mid-life; she calls it "the manly," in contrast to the more explicitly developmental terms "childhood," "youth," and "old age." (Indeed, in the course of the entire poem the words "middle age" appear only once.) Still, Bradstreet does explicitly address the substance of experience in this stage of life, as other writers generally did not.[14]

And what does she say about it? First, middle age is "graver" and "more staid" than the preceding stages—and carries with it a persistent concern for "a good report" (that is, for public reputation). But more central, at least in this description, is a certain attitude of life: an active, *productive* stance toward the wider social field. The middle-aged man is said to "grasp the world together . . . with both hands." He provides for his children, and also for his other kin. He undertakes important social responsibilities: he "feeds the poor," he administers justice, and (depending on his particular rank and station) he provides leadership and/or willing service in the life of his community. This is the positive side of the matter; Bradstreet also dwells at some length on the faults of middle age. The latter include *ambition* ("sometimes vainglory is the only bait whereby my empty soul is lured and caught"), *envy* ("then kings must be deposed or put to flight; I must possess that throne which was their right"), and an exaggerated concern with *wealth* ("his golden god's in his purse"). There is, evidently, a common thread here, running through both the virtues and the vices attributed to middle age: they have to do with the attainment and exercise of power. Mid-life is represented as a time of maximum capacity and opportunity; the question is whether this will be used for good or ill.

Bradstreet is, of course, a single voice from early America, but her account of middle age seems to match the actual experience of her neighbors and peers. The period of life from thirty to sixty was indeed a time for the exercise of power, for the realization of inborn potentials, for care and responsibility toward others. To be sure, those thirty years made in themselves a long span and, must in any individual life, have traced numerous turnings (and peaks and valleys as well). Still, it should be possible to construct an *average* profile around the themes of "ascent" and attainment, as noted earlier.

Typically, a man of thirty was just beginning his own family-life. Married a few years before, he had left the home of his parents, and with his bride had set up housekeeping in a separate residence.[15] (The residence itself might be newly built—and thus a symbol not only of change but of generative power.[16]) A child had been conceived within scant weeks or months of the wedding, and was now a living presence; others would arrive at regular intervals thereafter.[17] Each child brought undoubted responsibilities (for the parent), but was potentially a labor-source, a "comfort" in time of distress, and a guarantor of personal continuity into the future.

The same man was also beginning his career as an independent farmer or tradesman. He had his own land to look after (though without, in some cases, having received full legal title to it).[18] And he had his own livelihood to make. No doubt he continued to exchange some goods and services with his parents, siblings, and in-laws; but, for the largest part, he was on his own. The sum of his property was modest indeed, but (barring the "frowns" of providence, or the effects of his own incompetence) it would increase markedly in the years to come. His progress in this respect can be precisely measured through the tax and probate records of his community. Between thirty and forty-five the value of all his holdings was likely to increase two- or three-fold; and it would continue to increase into, or through, his fifties. This was the apex, however; after sixty, upon entering what he and his fellows called "old age," his wealth would gradually decline.[19] By this time his eldest children had themselves become adult and required from him their own "portions" of land and other property. (See Appendix, Table 1.)

A similar curve can be drawn for his participation in public life. Like most other males in early New England he must be "made free" by official order of the local magistrates—that is, empowered to vote and hold office. Typically, this would happen within a few years on either side of his thirtieth birthday. Soon thereafter he would begin to occupy minor positions in his community (fenceviewer, highway surveyor, constable, and the like). As he reached his forties and fifties he might, if all went well, be chosen for important town committees or even for the highest local office of all—that of selectman. In his sixties he would appear less often in the lists of officeholders, in his seventies not at all. To be sure, this picture would apply most fully if he began and remained in the "middling ranks" of his community. If his status should fall much below the average, he was unlikely to be elected to any office; and if he belonged to the local elite, he might serve in the highest positions irrespective of age.

There was one more public setting in which his progress could be traced. New Englanders made a notably litigious folk; they appeared in court time after time, in one cause or another.[20] Our hypothetical Everyman would be there as a plaintiff, as a defendant, and (most often) as a witness. His presence in the latter role was not age-determined; he was just as likely to witness in his thirties as in his forties or fifties. (See Appendix, Table 2.) However, in other respects his record in court displayed considerable variance according to age. How liable was he to be "presented" for offenses against community law and/or mores? A graph of the probabilities is highest for the young-adult years (25–29), still high but declining in the decade of the thirties, lower and essentially flat for the years from forty through sixty-five, and declining again to its lowest point of all in old age. The rate of such presentments drops by half between twenty and forty, and is more than halved again between sixty-five and seventy-five. (See Appendix, Table 3 and Figure 1.) What, on the other hand, of his initiatives as *plaintiff* against those by whom he considered himself wronged? The curve is largely reversed: that is, low at the start, rising steeply through the thirties, highest in the forties and fifties, declining slowly in the sixties, and lowest of all after seventy. (See Appendix, Table 4 and Figure 2.)

These quantitative measures may look rather sparse on the printed page, but they do tell us something of the underlying rhythms of aging in that time and place. Thus young men—those in transit to full adult status—were particularly liable to run afoul of the law. Perhaps the transition was itself problematic, with its blend of old obligations (to the family of origin) and new responsibilities (for the family formed by marriage and procreation). Scholars have noticed a pattern of tension, and occasionally of overt conflict, around the transfer of property from fathers to grown sons in early New England.[21] And perhaps this, in turn, expressed innerlife conflicts over the passage from dependence to autonomy, from *care by*, to *caring for*, a circle of significant others. One particular branch of legal procedure—witchcraft prosecutions—may be taken to support this hypothesis. For such prosecutions cast young men not as defendants, but rather—in highly disproportionate numbers—as "victims" and complainants. Their livestock would suffer sudden illness and death, their household properties would disappear, their skills would diminish—all seemingly without reference to "natural causes." These complaints expressed a sense of vulnerability that was quite specific to the (20s and 30s) age-group.[22]

The pattern for plaintiffs—in civil actions—also bears consideration. To bring a neighbor to account for non-payment of debt or other breach of contract, for defamation, for damage done to personal property, was to assert one's own interest in a forceful (and public) way. And perhaps it is not surprising that such self-assertion came most easily to men in their forties and fifties (as compared with other parts of the life course). Thus interpreted, suits-at-law are a measure of the psychology of men in full possession of their powers and properties. Situated at the apex of life's up-and-down path, such men experienced maximum wealth, maximum public influence and visibility, maximum progeny, and also—we are here suggesting—maximum confidence.

And what of women, viewed in the same general terms? Along with men—perhaps *more* than men—women in mid-life knew the power of parenthood. A New England "goodwife" in her forties and fifties would likely preside over a household of ten or a dozen; not for her the "empty nest," either in actuality or in prospect— quite the contrary. Her relation to her daughters (and maidservants "bound over" from elsewhere) was especially encompassing; to them she stood as ruler, tutor, example, protectress, all in one. Her duties were manifold, her skills indispensable. Gardening, cooking, dairying, textile production: thus the leading forms of specifically female labor, each one in itself subdivided into numerous smaller operations.[23] "In the agricultural towns and villages of early New England," one scholar has written, "women were the ones whose knowledge guided the key transformation of nature into culture." Men "broke" the raw materials of Nature (e.g. by cultivating and harvesting a variety of animal and vegetable products), while women "improved" these same materials (through processing and preparing them for human consumption).[24]

This sounds like a "domestic" definition of women's lives; and so indeed it was. Yet with some regularity women stepped out into roles and activities beyond the home-hearth. They entered the local economy, especially in marketing herbs, cheeses, cloth, and other products of their own hand. They practiced a traditional "doctoring," sometimes in direct competition with male physicians. (One branch of medicine belonged exclusively to women: childbirth was for female hands and eyes only, with midwives officially in charge.*)[25] They assisted their menfolk in running taverns, inns, and small shops.[26] At least occasionally they went into court as "attorneys" for temporarily absent husbands. There is enough

---

*It is pertinent, in the present context, to inquire about the age of New England midwives. There is no empirical research directed specifically to this question, but impressionistic evidence points to the middle years (and later). Perhaps a woman would not be accepted as a midwife until she herself had borne a full complement of children.

evidence here for one scholar to have invented the term "deputy husband" as a way of describing the full range of women's responsibilities in early New England.[27]

We say "women's responsibilities" without reference to age; but, in fact, we mean women in mid-life most especially. The scope and complexity of all this grew with the passing of each year after marriage, with the arrival of each new child, with the steady augmentation of family property and influence. And the sum of it was greatest in a woman's fifth and sixth decades, when her household reached its maximum size. Too, her status followed that of her husband in the community at large; the "deputy" rose along with her leader.

For all that, it would be a mistake to assume a congruence of developmental process in male and female—whether *then* or *now*. And there is at least indirect evidence that the middle years, particularly the fifth and sixth decades, brought distress as well as reward to New England women. There is, for example, the evidence of their activity in the courts. Women accused of criminal conduct (theft, blasphemy, assault, murder) seem to have come to a disproportionate extent from the mid-life cohort.[28] Conversely, women-as-deponents (in court cases involving others) included disproportionately few members of the same cohort. (See Appendix, Table 5.) The explanation for these discrepancies is obscure, but their surface meaning seems clear. Women in the middle years were more likely (than women of other ages) to run afoul of official norms, and less likely to provide support for those norms. In some broad sense, therefore, they faced away from the moral center of their community.*

More emphatic on the same point is the evidence of witchcraft proceedings. Indicted witches were largely women over the age of 40; their careers, when examined over time, typically reveal a pattern of suspicion against them beginning in their forties or fifties. Indeed, this pattern is so strong that one might almost speak of witchcraft—or, rather, the development of a reputation for witchcraft—as an age-specific event.[29] Furthermore: the *accusers* of witches, those persons who pressed for prosecution and came

---

*An important study of "deviance" in early New England calls attention to the leadership role of women in three critical episodes: the Antinomian Controversy of the 1630s, the Quaker insurgency of the late 1650s, and the witchcraft panic of the early 1690s. And, evidently, most of the women involved were of middle age (or older). See Kai T. Erikson, *Wayward Puritans: A Study in the Sociology of Deviance* (New York, 1966).

into court as witnesses and complainants, were drawn largely from the same age-group. Of course, they saw themselves not as accusers but as victims, as targets, as sufferers of the most extreme sort. In making their complaints they expressed a sense of vulnerability that was, again, specific to their life-stage.[30]

The witchcraft records hint, finally, at the sources of this recurrent pattern. Of what did the "victims" chiefly complain? Of physical illness and injury (not explained by "natural" causes) both in their own persons and in their offspring. And how, for that matter, did they characterize the accused? As malign, grasping figures who showed some particular interest in—and animus toward—infants and young children. Indeed, the entire panoply of witchcraft belief spotlighted the theme of *maternal* function. Witches were alleged to suckle their own "familiar" creatures (mice, cats, puppies, "imps"), and New England courts regularly ordered that witch-suspects be searched for the "marks" or "tits" which Satan had bestowed on them for precisely this purpose. Witches were frequently charged with interfering in the breast-feeding practice of ordinary mothers. Witches were, in reality, often deviant in their own child-bearing experience (having fewer children than the community average or none at all): hence, presumably, their envious interest in the children of others.[31]

There is a melody here of impressive consistency. In one way or another witchcraft cases announced (what we would call) the "menopausal" status of the leading participants. And in early New England, as throughout the pre-modern world, menopause had a clearer, sharper meaning than is the case today. It meant, of course, the end not only of reproductive capacity, but also of reproductive *experience*: women who had been bearing children for up to two decades would bear no more.[32] And how would they react to the change? The question is easier to ask than to answer, but we may reasonably imagine a mix of factors. There must, for one thing, have been feelings of relief and release. Child-bearing was physically burdensome and dangerous (as New Englanders of both sexes well understood); to arrive at menopause was, quite simply, to improve one's life chances.[33] Still, the witchcraft records—and historical intuition—suggest that there may have been another side. Child-bearing was, after all, a primary power of women, something which defined their role and nature in the most profound way. "Be fruitful and replenish the earth," commanded the Bible; and Puritan women were unusually successful in their response. Thus when the power of fruitfulness left them, they might well feel unsettled—or, in the language of witchcraft cases, "sorely afflicted." Perhaps no other point in a woman's life brought as sharp a turning as the end of child-bearing. (The start of child-bearing, by contrast, was prepared by the months of a first pregnancy, and before that by years of formal and informal socialization all pointing toward motherhood.)

This line of discussion yields at the end a picture of women's mid-life experience somewhat at variance with the typical pattern for New England men. In both groups, we may still affirm, the years between thirty and sixty produced maximum access to power and responsibility. Whether male or female, a New Englander would during these years climb to a kind of personal summit—and would stand there at least briefly before beginning (after age sixty) a marked descent. Recall Anne Bradstreet's characterization of middle age as "the manly" time. This was not, of course, a calculated reference to gender; she meant, instead, "manlike" or "fully human." But historical hindsight adds a touch of irony to her use of the term. Middle age was, in one important sense, more "manly" for men than womanly for women.

IV

Middle age is, even now, sometimes described as "the prime of life." Moreover, power and wealth attach almost as much as ever to people in their middle years. The effects of historical change should not, therefore, be construed as total ones. In a sense they involve certain *addenda* to mid-life experience: a new stage-consciousness, and deep feelings of unsettledness amounting in some cases to "crisis" (and evident most especially in men). Still, change generally holds more interest for historians than continuity. And change there is here, change aplenty.

In concluding, it may be appropriate to invoke the larger issue of life cycles in history. One reflection immediately suggests itself. There is a considerable body of recent scholarship which downgrades the salience of childhood and adolescence as developmental categories in the past.[34] And now we have surmised that middle age, too, meant rather less *then* than it does at present. The trend seems all in the same direction—though old age (considered in a subsequent chapter) might well emerge as a significant exception.

Keith Thomas has suggested that increasing age stratification is one of the truly big themes of early modern history. Since at least the seventeenth century the grouping of populations by age has become more and more formal and precise.[35] The development of age-specific roles and activities provides the front line of the evidence here; but the historical elaboration of "life cycle" experience also seems highly pertinent. Indeed, age-grouping and stage-definition might well be regarded as the outside and inside of the same trend. Or—in still other words—these are complementary ways of seeing historical change, the one sociological, the other essentially *psy*chological.

There are related considerations that might well repay some careful study. For example, it appears that precision as to chronological age is a relatively modern phenomenon. In early New England many people seem not to have known—or at least not to have cared—exactly how old they were.[36] (Interestingly, this pattern has some particular association with mid-life; the evidence "breaks" sharply, in the direction of much greater imprecision, for persons past thirty.) These seemingly trivial details point, in fact, to profound changes in human experience. *We* know how old we are at any given point; indeed, age is an aspect of our very sense of self. And this is, in effect, still another way of saying that our passage through life is much more highly codified—and self-conscious—than was normally the case for our forebears.

It is left to ask how long this process may continue. Will we eventually mark life's changes not simply in terms of advancing years, but of months and days? And will we discover additional "stages" within which to encompass the varieties of our experience? There are already some signs pointing in that direction. Kenneth Keniston has recently tried to describe a new life-stage, post-adolescent but pre-adult, which he calls, simply, "youth."[37] Daniel J. Levinson has carved contemporary middle age into some ten sub-phases (alternating "structure-building periods" with "transitions").[38] And Bernice Neugarten, investigating the farther end of the life cycle, has proposed two subdivisions of old age as typically experienced in modern America, the "young-old" and the "old-old."[39]

But these trend-lines carry us well beyond the territory of the historian. It is hard enough, in life-cycle studies, to discern the past—and harder still to divine the future.

NOTES

1. M. W. Lear, "Is There a Male Menopause?" *The New York Times Magazine* (January 28, 1973).

2. Among the best-known of the scientific studies are: Daniel J. Levinson *et al.*, *The Seasons of a Man's Life* (New York, 1978); George Vaillant, *Adapta-*

tion to Life (Boston, 1977); Roger Gould, "The Phases of Adult Life: A Study in Developmental Psychology," *American Journal of Psychiatry*, CXXIX (1972), 521–31; and the numerous writings of Bernice L. Neugarten and her associates. For an excellent review of these and other studies, see Orville G. Brim, Jr., "Theories of the Male Mid-Life Crisis," *The Counseling Psychologist*, VI (1976), 2–9. Among "popular" works the most popular by far has been Gail Sheehy, *Passages: Predictable Crises of Adult Life* (New York, 1976).

3. Levinson, *The Seasons of a Man's Life*, 322.

4. Cotton Mather, *Addresses to Old Men . . .* (Boston, 1690).

5. See Elliot Jacques, "Death and the Mid-Life Crisis," *International Journal of Psychoanalysis*, XLVI (1965), 502–14; and J. M. A. Munnichs, *Old Age and Finitude: A Contribution to Psychogerontology* (New York, 1966).

6. Commonplace Book of Joseph Green, in *Remarkable Providences*, John Demos, ed. (New York, 1972), 350.

7. Alfred Kinsey *et al.*, *Sexual Behavior in the Human Male* (Philadelphia, 1948) and *Sexual Behavior in the Human Female* (Philadelphia, 1953); William H. Masters and Virginia E. Johnson, *Human Sexual Response* (Boston, 1966).

8. See, for example, Norman E. Zinberg and Irving Kaufman, *Normal Psychology of the Aging Process* (New York, 1963); and Helmut J. Ruebsatt and Raymond Hull, *The Male Climacteric* (New York, 1975).

9. See Edmund S. Morgan, "The Puritans and Sex," *New England Quarterly*, XV (1942), 591–607; and John Demos, *A Little Commonwealth: Family Life in Plymouth Colony* (New York, 1970), 152–59.

10. See David Gutmann, "The Cross-Cultural Perspective: Notes Toward a Comparative Psychology of Aging," in *Handbook of the Psychology of Aging*, James E. Birren and K. Warner Schaie, eds. (New York, 1977), 302–26.

11. *Ibid.* See also David Gutmann, "The Country of Old Men: Cross-Cultural Studies in the Psychology of Later Life," in W. Donahue, ed., *Occasional Papers in Gerontology* (Ann Arbor, Mich., 1969).

12. See Jacques, "Death and the Mid-Life Crisis," 506–7, and the essays included in *Middle Age and Aging: A Reader in Social Psychology*, Bernice E. Neugarten, ed. (Chicago, 1968).

13. See the preceding essay in this volume, Chapter Five.

14. Anne Bradstreet, "The Four Ages of Man," in *The Works of Anne Bradstreet*, Jeannine Hensley, ed. (Cambridge, Mass., 1967), 51–63.

15. For example: the average age at first marriage, for men in Plymouth Colony during the seventeenth century, was approximately 27 years. See Demos, *A Little Commonwealth*, 193. Similar figures have been obtained for seventeenth-century Andover and Ipswich, Massachusetts; see Philip J. Greven, Jr., *Four Generations: Population, Land, and Family in Colonial Andover, Massachusetts* (Ithaca, N.Y., 1970) and Susan Norton, "Population Growth in Colonial America: A Study of Ipswich, Mass.," *Population Studies*, XXV (1971), 433–52.

16. The parents of a couple planning marriage would usually undertake a variety of financial obligations, formally recorded in a legal contract and sometimes including explicit provision "to build a comfortable dwelling house" (for the newlyweds). See for example, the "Marriage Contract for Jacob Mygatt and Sarah Whiting (Hartford, Connecticut, 1654)," in *Remarkable Providences*, Demos, ed., 145–47.

17. On birth spacings, see Demos, *A Little Commonwealth*, 68, 133.

18. The complex question of the transmission of land from one generation to the next is explored at length in Greven, *Four Generations* and Demos, *A Little Commonwealth* (especially 164–70).

19. See the following essay in this volume, Chapter Seven.

20. Virtually every scholar who has examined early New England court records has come away with this impression. As yet there is no published work presenting the matter of litigiousness in quantitative terms; but my own research, currently in progress, suggests that an average person would appear as a defendant about once in four years, as a plaintiff slightly less often, and as a deponent perhaps once every other year. This, at least, is the picture for the quarterly court of one county (Essex) in Massachusetts. If lesser courts (e.g., those for individual towns, with single magistrates presiding) were considered too, the overall litigation-rate would rise even higher.

21. See especially Greven, *Four Generations*.

22. This pattern is treated at length in John Demos, *Entertaining Satan: Witchcraft and the Culture of Early New England* (New York, 1982), 144–47, 153–54, 156–57.

23. The best single study of the lives of early New England women is Laurel Thatcher Ulrich, *Goodwives: Image and Reality in the Lives of Women in Northern New England, 1650–1750* (New York, 1981).

24. See Robert Blair St. George, "'Set Thine House in Order': The Domestication of the Yeomanry in Seventeenth-Century New England," in *New England Begins: The Seventeenth Century*, 3 vols. (Boston, 1982), II, 170.

25. On childbirth in early New England, see Richard Wertz and Dorothy Wertz, *Lying-In: The History of Childbirth in America* (new York, 1977), chs. 1–2.

26. See Elizabeth Anthony Dexter, *Colonial Women of Affairs*, and Ulrich, *Goodwives*.

27. *Ibid.*

28. This is, admittedly, more an impression then an empirically demonstrated fact. Gathering evidence about the age of women criminals is extraordinarily difficult, but my own small sample (some 30 individuals "presented" to the quarterly court of Essex County) yields the following result: approximately one-quarter in the age-group 20–39; two thirds in the age-group 40–59; the remainder over 60. (Note that these findings *exclude* sexual offenses, chiefly fornication by very young women.)

29. Among accused "witches" in seventeenth-century New England, at least 80 percent came under suspicion when they were between 40 and 59 years of age. See Demos, *Entertaining Satan*, 66–67.

30. *Ibid.*, 144–47, 154–56.

31. For a much fuller treatment of witchcraft and its "maternal" associations, see ibid., 72–73, 179–81, 198, 204–5.

32. Most New England women began child-bearing within a year of their marriage (typically in their early or middle twenties) and ended in their early forties. One careful study of this matter, based on records from the town of Rowley, Massachusetts, showed 65 percent of women bearing their last child at ages over 40, and 10 percent at over 45. (See Patricia Trainer O'Malley, "Rowley, Massachusetts, 1639–1730: Dissent, Division, and Delimitation in a Colonial Town," unpublished Ph.D. dissertation, Boston College, 1975, p. 211.)

33. Thus life expectation was significantly lower for women than for men at

all ages below 50, but higher thereafter. See the figures for seventeenth-century Plymouth, in Demos, *A Little Commonwealth*, 192.

34. The seminal work here was Philippe Ariès, *Centuries of Childhood: A Social History of Family Life*, Robert Baldick, trans. (New York, 1962). See also Demos, *A Little Commonwealth*, 57–58, 134–44.

35. Keith Thomas, "Age and Authority in Early Modern England," *Proceedings of the British Academy, LXII* (1976), 1–46.

36. Once again I must refer to parts of my own research, unpublished and "in progress." Nearly 2000 residents of Essex County, Massachusetts, stated their age when giving evidence before the quarterly courts; many of them did so more than once. When analyzed *en masse*, these statements show very strongly the effect known to demographers as "age-heaping," that is, a tendency to "round off" to intervals of ten (30, 40, 50, and so on). In addition, comparison of age-reports given by the same individual (on different occasions) shows frequent inconsistency—usually within a range of one to three years.

37. Kenneth Keniston, "Youth as a New Stage of Life," in *Youth and Dissent* (New York, 1971), 3–21.

38. Levinson, *The Seasons of a Man's Life*.

39. Bernice L. Neugarten and Gunhild O. Hagestad, "Age and the Life Course," in Robert H. Binstock and Ethel Shanas, eds., *Handbook of Aging and the Social Sciences* (New York, 1976), 46 ff.

APPENDIX: TABLES AND FIGURES

TABLE 1. Correlations of Age and Wealth in Colonial New England

Age Groups	Hampton Tax List (1653)	Hampton Tax List (1680)	Easthampton Tax List (1675)	Inventories, Five Towns, 17th Century*	Inventories, Lyme, Conn., 1676–1776
20–29	43.7**	86.7**	41.3**	£112	£ 72
30–39	41.2	61.3	37.0	134	224
40–49	36.1	45.8	17.6	431	442
50–59	13.9	28.1	9.0	555	517
60–69	22.0	39.0	16.4	434	335
70–79	55.0	42.3	. . .	308	262
80–89	. . .	94.5	. . .	286	. . .
N	73	123	54	196	. . .

*Sample includes estate inventories for men whose age of death can be determined from the following towns: Hampton (N.H.), Springfield (Mass.), Northampton (Mass.), Wethersfield (Conn.), Easthampton, L.I. (part of Conn.). Average values given in pounds (£).

**Average rank, relative to all other taxpayers on complete list.

SOURCES: Local records and genealogies from the towns involved, family reconstitution by the author, and Jackson Turner Main, "The Economic and Social Structure of Early Lyme," in George J. Willauer, Jr., ed., *A Lyme Miscellany* (Middletown, Conn., 1977).

TABLE 2. Age of Witnesses (Male)

Ages	As Percentage of Total Population	As Percentage of All Witnesses
20–29	29	28.5
30–39	25	24.8
40–49	20.5	19.5
50–59	15	13.9
60–69	6.5	9.2
70–79	4	3.9

NOTE: Middle column shows size of each ten-year cohort as percentage of total adult population (i.e. all males over 20 years) in 17th-century New England towns. Right-hand column shows size of each cohort as percentage of all adult male witnesses deposing in the quarterly courts of Essex County, Massachusetts, during the period 1636–82. When directly compared the figures in these two columns seem very similar, except for the cohort 60–69 (in which witnesses are more numerous than their presence in the general population might predict).

SOURCES: Figures in middle column are averages of findings obtained from demographic reconstruction of numerous New England communities (by the author). Figures in right-hand column are based on age-reports given by all male witnesses (over 20) recorded in the files of the quarterly courts of Essex County, Massachusetts. See *Records and Files of the Quarterly Courts of Essex County, Massachusetts*, 8 vols. (Salem, Mass., 1916–20), *passim.*

TABLE 3. Age of Offenders (Male)

Ages	Years at Risk	Adjusted Years at Risk	Offenses	Rate per 100
25–29	136	218	14	6.42
30–34	299	343	20	5.83
35–39	386	411	18	4.38
40–44	435	455	14	3.08
45–49	474	. . .	15	3.16
50–54	480	. . .	15	3.13
55–59	430	. . .	14	3.26
60–64	358	. . .	12	3.35
65–69	253	. . .	6	2.37
70–74	158	. . .	2	1.27
75–79	84	. . .	1	1.19

NOTE: Results are based on the careers of 100 adult male residents of Essex County, Massachusetts during the period 1636–82. Each man in the sample was present for at least 25 years. Court records were examined, covering every year of such presence. Totals of man/years for the entire sample, as distributed in five-year cohorts, are shown in second column from left ("Years at Risk"). Because a considerable portion of men in the sample were present in Essex County before the date of their first appearance in the court records, an "adjustment" has been made in the "Years at Risk" figures. (The adjustment adds, to each cohort under 45, half the difference between it and the next older cohort.) This procedure yields the figures in the third column from left ("Adjusted Years at Risk"). Fourth column from left ("Offenses") shows totals of cases in which a man within the given age-cohorts was "presented" before the county court. (Offenses include theft, breach of the peace, assault, public drunkenness, sexual misdemeanors, and other violations of county law.) Last column ("Rate per 100") measures figures in fourth column against those in third. In short: among men between the ages of 25 and 29 years, there were 6.42 offenses per 100 man/years; for men between 30 and 34, 5.83 per 100; and so on.

SOURCES: *Records and Files of the Quarterly Courts of Essex County, Massachusetts,* 8 vols. (Salem, Mass., 1916–20), *passim.*

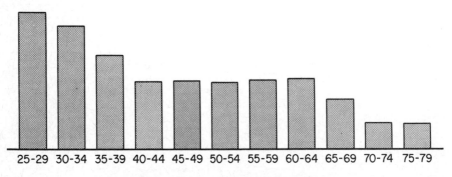

FIGURE 1. Age of Offenders (Male)
(Expressed as rate per 100; see Table 3)

TABLE 4. Age of Plaintiffs (Male)

Ages	Years at Risk	Adjusted Years at Risk	Years as Plaintiff	Rate per 100
20–24	55	96	4	4.16
25–29	136	218	9	4.13
30–34	299	343	17	4.96
35–39	386	411	35	8.52
40–44	435	455	42	9.23
45–49	474	. . .	62	13.08
50–54	480	. . .	52	10.83
55–59	430	. . .	45	10.47
60–64	358	. . .	34	9.50
65–69	253	. . .	21	8.30
70–74	158	. . .	6	3.80
75–79	84	. . .	1	1.19

NOTE: Results in the table are based on the careers of 100 adult male residents of Essex County, Massachusetts during the period 1636–82. Each man in the sample was present for at least 25 years. Court records were examined, covering every year of such presence. Each man/year was scored either "positive" (one or more cases filed as plaintiff) or "negative" (no cases as plaintiff). Totals of man/years for the entire sample, as distributed in five-year cohorts, are shown in second column from left ("Years at Risk"). Because a considerable portion of men in the sample were present in Essex County before the date of their first appearance in the court records, an "adjustment" has been made in the "Years at Risk" figures. (The adjustment adds, to each cohort under 45, half the difference between it and the next older cohort.) This yields the figures in the third column from left ("Adjusted Years at Risk"). Fourth column from left ("Years as Plaintiff") shows totals of "positive" years (as defined above). Last column ("Rate per 100") measures figures in fourth column against those in third. In short: for men between the ages of 20 and 24 years, the chances of appearing in court as a plaintiff (in any given year) were 4.16 in 100; for men between 25 and 29, 4.13 in 100; and so on.

SOURCES: *Records and Files of the Quarterly Courts of Essex County, Massachusetts,* 8 vols. (Salem, Mass., 1916–20), *passim.*

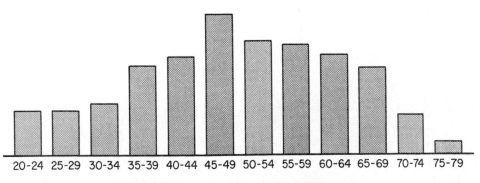

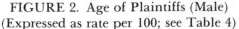

FIGURE 2. Age of Plaintiffs (Male)
(Expressed as rate per 100; see Table 4)

TABLE 5. Age of Witnesses (Female)

Ages	As Percentage of Total Population	As Percentage of All Witnesses
20–29	33	33.5
30–39	27.5	27.4
40–49	21	15.5
50–59	12	12.9
60–69	4.5	6.5
70–79	2	4.2

NOTE: Middle column shows size of each ten-year cohort as percentage of total adult population (i.e. all females over 20 years) in 17-century New England towns. Right-hand column shows size of each cohort as percentage of all adult female witnesses deposing in the quarterly courts of Essex County, Massachusetts, during the period 1636–82. When directly compared the figures in these two columns seem similar for the 20–29, 30–39, and 50–59 cohorts. However, witnesses are much *under*-represented (relative to the total population) for the 40–49 cohort, and somewhat over-represented for the 60–69 and 70–79 cohorts.

SOURCES: Figures in the middle column are averages of findings obtained from demographic reconstruction of numerous New England communities (by the author). Figures in the right-hand column are based on age-reports given by all female witnesses (over 20) recorded in the files of the quarterly courts of Essex County, Massachusetts. See *Records and Files of the Quarterly Courts of Essex County, Massachusetts*, 8 vols. (Salem, Mass., 1916–20), *passim*.

CHAPTER VII

Old Age in Early New England

Aging became a "hot" topic with government and foundation planners during the decade of the 1970s. Abruptly, coffers opened; rhetoric resounded; new agencies and institutes sprang to life; programs and policies proliferated at a great rate. As usual, denizens of the Academy caught the scent and went after a piece of the action. Social scientists of various stripe got there first, but historians were not far behind.

In fact, official largesse *nicely coincided with trends and tendencies internal to the discipline. By this time the life course had become fully legitimated as a credible target for historical study, and scholars were flocking to it even without special inducement. Initially, however, their sights had been set on the early parts of the life course; experience after about age twenty remained a largely uncharted territory. Old age waited, then, as a logical "next step."*

A single scholarly group seized the initiative, and set the otherwise muddled ranks in marching order. Case Western Reserve University was its site, "Human Values and Aging" the name given to its "project." As with child abuse (see above, Chapter Four), there was a feeling that "a humanistic perspective . . . [might] . . . enhance . . . present understanding of a . . . social problem of growing proportions."

Again, I was one of the "humanist" draftees. And I welcomed the call for reasons entirely my own. Years before, I had written a small piece on the history of old age (a chapter in a book on early American family life) which seemed in retrospect painfully inade-

This essay was first published in John Demos and Sarane Spence Boocock, eds., *Turning Points: Historical and Sociological Essays on the Family* (Chicago, 1978), 248–87.

quate. Now I would have that rarest of scholarly opportunities, a second chance. This time I would do it right.

The reader will judge the extent of my success. But he/she should be forewarned on two or three important counts. This essay differs from all others in the volume: (1) in its length; (2) in its narrow focus (a single time and place); and (3) in its extended presentation of both research method and empirical detail. Though it could not properly be characterized as "technical," it makes a different sort of reading.

Different, but much more representative of the usual bedrock labor in family history. That, indeed, is my rationale for including it here: behold the digger at his daily task!

I

Within the past decade, scholars have come to appreciate the significance of age as a determinant of historical experience. At first their interest was directed largely to childhood, then to adolescence, and now—in appropriate sequence—to the later parts of the life course. For historians, among others, old age is a time whose idea has come.

The overall topography of the field has been charted in a bold and brilliant book by David Hackett Fischer entitled *Growing Old in America*.[1] In the span of little more than two hundred pages, Fischer has ranged through the entire history of old age in the United States, and added some brief but vivid forays across the oceans to Europe and to Asia, back down the centuries to ancient and medieval times, and even (in a closing chapter) through the looking glass into the future. At the heart of his study, Fischer has fashioned an interpretive scheme—a "model" of the history of age relations—which seems likely to influence research in this field for years to come. Other scholars will now wish to cover particular sections of the larger territory more slowly and with a somewhat sharper focus.

Some such justification is needed for the present essay, for the concern here is with a small and rather remote corner of the American past. Little attention is given to issues of chronological development and change. Instead, the main goal is to present a rounded view of old age in one particular setting; the materials are explored—and where possible, connected—on a "horizontal" rather than a "vertical" basis.

Even within this modest framework of intentions, problems arise and limits must be accepted. There are, for one thing, severe problems of evidence. The historical record of early New England includes a substantial corpus of prescriptive statements on old age, chiefly sermons and essays by leading clergymen. There is, however, a dearth of evidence directly reflecting behavior by and toward the elderly. Thus the scholar is obliged to pursue indirect methods: for example, demographic reconstruction, the analysis of legal materials, and assorted forms of collective biography. Some important questions inevitably slip by unanswered. The most glaring *lacunae* in the present case involve the experience of New England women. Quite possibly male aging and female aging were significantly different, but the data seem too meager to permit even an opinion about this, let alone any solid conclusions.*

It may be unfair to hold the evidence entirely to blame for such deficiencies. After all, historical evidence responds to—and, in a sense, is created by—the asking of particular questions. And in any "new field" the questions themselves are problematic. It thus seems prudent to specify in advance the questions to be asked here. First, we shall consider how old age was conceived—that is, how it was thought about—in early New England. Second, we shall assess the elderly as a demographic presence: how many they were, and how they were situated in the larger population. Third, we will explore the social experience of old people: their work, their leisure, the nature of their power and prestige. The essay will close with some interpretive suggestions and a residual agenda of questions for the future.

II

Cotton Mather of Boston, in publishing a treatise entitled "The Old Man's Honour," included a frothy dedication to his elderly friend and mentor, Major John Richards. He wrote in part: "Were there nothing else to commend my regards for you, besides the Old Age, which your out-living of three-score winters has brought you to the border of, that were enough to give you a room in my esteem, and reverence, and veneration."[2] This passage helps to illumine two central questions pertaining to old age in early New

---

*However, for some discussion of *mid-life* experience among early New England women, see Chapter Six.

England: its chronological definition and its claims for attention in the culture at large. Each point merits detailed consideration.

The view of age sixty as a "border" appears in various written statements, and is implied in a different way by certain forms of legislative enactment. A clergyman noted "the considerable number of aged persons" in his congregation: "For there are many who have attained to three score, and such are everywhere accounted old men." A town voted to exempt older persons from particular civic obligations and established "60 years of age" as the official cutoff point. A provincial assembly set requirements of regular service in paramilitary units for all male inhabitants "between the ages of 16 and 60 years."[3]

Here, then, one finds an apparent consensus as to the beginning of old age. And yet its application in specific instances was far from precise. The requirement of military service called forth some especially revealing evidence. When a man wished to be "freed from training" (i.e. with his local militia unit), he was obliged to petition the courts for a personal waiver. Thus (1) "John Leigh, being about seventy years of age, [is] discharged from ordinary training"; (2) "Robert Kinsman, being above three score years of age and having the 'seattyca,' was freed from training"; (3) "John Cooly, being aged, and having fits whereby he falls, is freed from training"; (4) "William Lord of Salem, aged seventy-seven, [is] discharged from training, on account of age and many bodily infirmities."[4] The list of such actions could be lengthened indefinitely, but the larger point is immediately clear. The actual age invoked was often more (and occasionally less) than sixty, was sometimes omitted altogether, and was usually linked to the physical condition of the man in question. This implies a certain looseness or flexibility in the application of age norms and, moreover, a functional attitude toward the process of growing old. Aging was measured, in part, by numbers, but also by the survival (or decline) of inherent capacity. The tendency of our own time is much more exclusively formal: chronology is usually the decisive consideration (e.g. in retirement), while biology counts for relatively little.

Cotton Mather's dedication to Major Richards spoke of "esteem, and reverence, and veneration" as appropriate attitudes toward the aged. Here, too, he expressed a widespread cultural convention. "Honour old age": so it was written in the Scriptures and endlessly repeated in the sermon literature of early New England.

There were various ways to justify this prescription. In the first place, the elderly are wiser than other persons; their "counsel" should therefore carry disproportionate weight in civil and religious affairs. Their wisdom derived, in turn, from the sheer element of accumulated experience. When the "Pilgrim pastor," John Robinson, described the preferment due to older people, he noted in particular "their manifold advantages . . . for the getting of wisdom." When the poet Anne Bradstreet composed her verses on "The Four Ages of Life," she chose a special way of portraying the last one. She listed the numerous "private changes" and "various times of state" to which an old man of her era might bear witness, virtually defining him as a repository of experience. It is not easy for us today to recapture the strength of these associations; they seem insubstantial, or simply trite. Our culture has many ways of storing experience, most obviously in written documents. But in communities where literacy was less extensive, it was often other people—elderly ones—who provided a sense of contact with the past. To the extent that the past was honored, they were honored too.[5]

For so-called Puritans there was a special religious dimension to the accumulated experience of old age. The elderly were thought to have "a peculiar acquaintance with the Lord Jesus Christ." In fact, this acquaintance verged on likeness: "there is something of the image of God in age," an almost physical resemblance. God, after all, "is the Ancient of Days," and when His "majesty amd eternity are set forth in Scripture, it is with white hair." Hence "the fear of God and honoring the old man is commanded with the same breath and linked together in the same sentence." These brief quotations, stitched together from various New England sermons, reflect an idea widely noted in studies of premodern society. Elderly persons are literally and figuratively closer than others to God. They stand, as it were, near the boundary between the natural and supernatural worlds. Indeed, they are in a special position to mediate between these worlds—which explains, in part, the preferments they customarily enjoy.[6]

To restate these prescriptive standards for dealings with the aged is to consider only one side of a highly complicated picture. Did old people, in their actual behavior, justify the confidence associated with age? Did their particular traits and tendencies inspire feelings of "esteem and reverence and veneration"? What was the typical predispostion of the elderly in terms of what we now call character or personality? Such questions are also treated in the literature of early New England, and in ways that jar uncomfortably with the ritual exhortation to "honour old age."

Consider, for example, the opinion of William Bridge, some-time fellow of Harvard College and author of the earliest treatise on this subject published anywhere in the colonies. "Old age is a dry and barren ground," he begins. "The state of old age is a state of weakness and of much infirmity." Bridge proceeds to spell this out in great detail, separating the natural from the moral infirmi-ties. The moral ones make an especially long list. Older people are likely to be "too drowsy and remiss in the things of God, . . . too covetous and tenacious for the things of this world, . . . too timor-ous and fearful, . . . too touchy, peevish, angry, and forward, . . . very unteachable . . . [since] they think they know more than others, . . . hard to be pleased, and as hard to please others, . . . full of complaints of the present times, . . . [and] full of suspicions, and very apt to surmise, suspect, and fear the worst." Parts of this portrait show up in other writings from the period; the overall effect is certainly unflattering.[7]

The matter of "natural infirmities" in old age also received much attention. Cotton Mather, for one, and the aforementioned William Bridge, for another, catalogued the physical aspects of aging in somewhat gruesome detail. Here is a short sample:

> The sun, the light, the moon, and the stars begin to be dark-ened with you; that is, your parts are under a decay; your fancy, your judgment are failing you . . . Your hands now shake and shrink, and must lean upon a staff . . . Your thighs and legs now buckle under you . . . Your teeth grow weak and few, and are almost all rotted out . . . Your eyes become dim, and clouds disturb the visive powers in them . . . You become deaf and thick of hearing . . . You can't without some diffi-culty go up a pair of stairs, and are in danger of stumbling at every stone in the street . . . Your backs are so feeble that instead of carrying anything else they can scare bear them-selves . . .[8]

Loss, decline, decay: these were the central images. There is a note of distaste here, almost of repulsion, which betrays an important pre-conscious attitude.

And there is more. One feels in the sermon literature a sharp, even scornful quality, insofar as such literature is directed to the elderly themselves. Thus we find Cotton Mather, at the ripe age of twenty-nine, hectoring his older parishioners to repent their sins and threatening them with dreadful figures of Hell. At one point Mather enumerates six particular "virtues . . . which all old men should be studious of": sobriety, gravity, temperance, orthodoxy, charity, and patience. Yet in discussing these, one by one, he emphasizes not the beauty of the virtue itself but rather the ugliness of its associated vice. Sobriety, for example, is contrasted with drunkenness, and the terms which Mather chooses to frame the comparison express a certain relish. "For them that stagger with age, at the same time to stagger with drink; to see an old man reeling, spewing, stinking with the excesses of the tavern, 'tis too loathsome a thing to be mentioned without a very zealous destestation."[9]

Indeed, all the vices of the elderly seemed especially detestable—and *visible*. One more striking image, from the essay by William Bridge, will help to make the point: "When the leaves are off the trees, we see the birds' nests in the trees and bushes. Now in our old age our leaves are off, then therefore we may see these nests of sin, and lusts in our hearts and lives, which we saw not before, and so be sensible and repent of them." Among other things, in short, old age brought exposure; personal character was stripped of its protective covering and made to seem ridiculous, even contemptible.[10]

III

We cannot proceed much further in this investigation without bringing forward demographic considerations. We need to assess the actual presence of the elderly in the total population of colonial New England. How numerous was the "old" cohort in comparison with other age groups? What were the chances that individuals might survive to old age? To what degree was personal contact with elderly people a regular feature of life? Answers to these questions should permit us to decide whether there was something intrinsically "special" and exotic about old age within the larger frame of social experience.

Among modern-day Americans, the age group of persons sixty years old and more comprises some 15 percent of the total. We have no precisely comparable figures for early America (no figures for large populations), but assorted findings from local communities are helpful in establishing approximate trends and tendencies.

There follows a brief summary of such findings based on the study of five different New England towns.

1. In 1678 all male inhabitants (over the age of 16) of Newbury, Massachusetts, were required to take an oath of allegiance to the Crown; in the process their names and ages were recorded on a list which is still extant in the files of the Essex County Court. The total number of these oath-takers was 206. Twenty-eight of them were at least sixty years old; 11 were at least seventy. The entire male population of Newbury in this year (i.e. including those younger than sixteen as well) can be estimated at 420. Thus the over-sixty cohort represented about 6.7 percent of the whole. For the over-seventy category the figure is 2.6 percent.[11]

2. The age structure of the population of Windsor, Connecticut, has been analyzed for the years 1640 and 1686. In the former case 1.3 percent of the settled inhabitants were found to be sixty or older; in the latter, 4.1 percent.[12]

3. Hampton, New Hampshire, has been studied by equivalent methods and also for two different points in time. Careful reconstitution of Hampton families in the year 1656 yields a roster of some 356 local inhabitants. At least 14 of these people were more than sixty years old—or 4 percent of the entire list. More than two decades later (1680) the male adults resident in Hampton were "rated" for tax purposes. Their number was 126 overall; 120 of them can be assigned at least approximate ages. In addition, 90 women can be definitely associated with these men (i.e. as wives or widows), and the entire population of the town (including children) can be estimated at about 525. At least 33 of the adults (22 men, 11 women) were older than sixty at this time, which translates to 6.3 percent of the total.[13]

4. Still another investigation of this type has been made with the townspeople of Wethersfield, Connecticut, as of the year 1668. The results are: a total population of 413, an over-sixty cohort of 9 (or 2.2 %). If the latter figures seem very small, they need to be considered in the light of a temporary demographic anomaly. The Wethersfield citizenry of 1668 included a disproportionately large group of persons in their fifties (26 altogether). Thus the officially "old" cohort must, within a few years, have grown considerably.

5. There exists for the town of Bristol, Rhode Island, an actual (and quite unique) census of local inhabitants made in the year 1689. An absence of reliable vital records precludes any full analysis of age groups in this case; however, approximate ages can be assigned to the 68 adult women on the list. Within the latter group

only two may possibly have been as old as sixty, and even there the evidence is ambiguous. Another local census was made in Bristol nearly a century later (1774), and by then, it can be shown, the over-sixty cohort had risen to 5.6 percent of the total citizenry.[14]

These somewhat scattered findings must now be sorted so as to yield a more general conclusion. (See Table 1.) In five of eight instances (Newbury, 1678; Windsor, 1686; Hampton, 1656; Hampton, 1680; Bristol, 1774) the "old" cohort falls within a range of 4 to 7 percent of total population. The remaining three cases (Wethersfield, 1668; Windsor, 1640; Bristol, 1689) all produce significantly lower results. We have seen, however, that the first was subject to a quirk in the numbers themselves. And there were special circumstances affecting the other two as well. Windsor in 1640 and Bristol in 1689 were new communities, barely removed from a "wilderness" state. Evidently the settlement process was an affair of young people, or at least of those not old; for some years,

TABLE 1. Age Structures in New England Towns in the 17th and 18th Centuries

Town	Year	Persons over 60	Total Population	% over 60	Total Adults (over 20)	% over 60
Newbury, Mass	1678	28	420*	6.7	206**	13.5
Windsor, Conn	1640	..	...	1.3	...	4.1
Windsor, Conn	1686	..	...	4.1	...	8.5
Hampton, N.H. ...	1656	14	356	4.0	136	10.3
Hampton, N.H. ...	1680	33	525*	6.3	240	13.8
Wethersfield, Conn	1668	9	413	2.2	174	5.3
Bristol, R.I.	1689	2 (?)	...	...	68	2.9
Bristol, R.I.	1774	..	...	5.6	...	10.6

NOTE—Newbury, Mass (1678) includes males only; Bristol, R.I. (1689) includes females only.

*Estimate

**Over 16 years old; elsewhere adults means over 20.

SOURCES: *Records and Files of the Quarterly Courts of Essex County, Massachusetts*, 8 vols. (Salem, Mass., 1911–21), VII, 156–67; Linda Auwers Bissell, "Family, Friends and Neighbors: Social Interaction in Seventeenth-Century Windsor, Connecticut" (Ph.D. dissertation, Brandeis University, 1973); Town Book of Hampton (New Hampshire), 2 vols. (manuscript holdings, in Town Offices, Hampton, N.H.); Town Votes of Wethersfield, Connecticut (manuscript volume, in Connecticut State Library, Hartford, Conn.); John Demos, "Families in Colonial Bristol, R.I.: An Exercise in Historical Demography," *William and Mary Quarterly*, Third Series, XXV (1968), 40–57.

in such places, the age structure of inhabitants was foreshortened. Later, as the communities themselves aged, so, too, did their populations, with an "old" cohort gradually filling out at the farther end of the demographic spectrum. These changes belonged to a regular morphology of town growth and development.

We may therefore regard 4 to 7 percent as the likely portion of elderly people in established New England communities. Yet these findings need to be refined in one additional connection. With fertility limited only by natural constraints (e.g. menopause and the contraceptive effects of lactation), the birth rate was consistently high; hence colonial populations were everywhere skewed toward youth. Persons under twenty normally made up a majority of the whole, and the relative size of all older cohorts was diminished accordingly. Perhaps, then, we should measure the elderly in relation to other adults. In modern America the over-sixty age group is some 23 percent of the larger adult population (defined, for the moment, as all those who are at least twenty). In our five leading cases from the colonial era, the comparable figures are: 8.5 percent (Windsor, 1686), 10.3 percent (Hampton, 1656), 10.6 percent (Bristol, 1774), 13.5 percent (Newbury, 1678), 13.8 percent (Hampton, 1680). This is to say that the numerical presence of old people among adults generally was about half as large in colonial New England as it is now. The difference is certainly substantial, but it does not seem overwhelming.

We shall now alter our line of approach, translating our interest in age structures and "cohorts" into questions about personal expectation of life. To what extent, as individuals, did New Englanders actually survive to old age? Was this a prospect only for a very few, or was it something that many younger people might reasonably anticipate? The evidence is less full and less reliable than scholars might ideally wish, and the currently available results show some points of difference and disagreement. Still, most signs point strongly in one general direction, with implications for the study of aging that are truly profound.

The first substantial research on "survivorship" in early New England appeared about a decade ago. Its geographical foci were the Massachusetts towns of Andover and Ipswich and the colony of Plymouth. In each case local populations were scanned for information on age at death, and the results, taken together, showed a very considerable pattern of longevity. In all three settings the majority of recorded deaths occurred among persons distinctly "old." (Indeed, at both Plymouth and Andover the most common age-decade for mortality was the seventies.) Given what is known of other pre-modern populations, these figures seemed quite incredible, and some scholars found them literally so.

Indeed, the Plymouth, Andover, and Ipswich studies displayed a number of methodological shortcomings. Particularly troublesome was the question of bias in the various sample populations. (Were short-lived persons less likely than others to find their way into the records?) New studies were called for, which would deal with this possibility more effectively; the returns are just now coming in. One recent investigation has canvassed all deponents in the quarterly courts of Middlesex County, Massachusetts, during the period 1661–75. By comparing age at the time of witnessing (usually recorded in the deposition itself) with age at death (where known), it was possible to construct a "life table" for this particular group. At birth 44.5 percent of the population might expect to live to age sixty or more, and 20.8 percent to at least seventy. Among those who survived to age twenty the figures rose to 54.9 percent and 34.6 percent respectively. Mean life expectation among the twenty-year-olds was an additional 40.5 years.[15] (See Table 2.)

An alternative strategy is to analyze the entire population of particular town-communities at given points in time. These populations can, with only marginal error, be reconstructed from local

TABLE 2. Survivorship in 17th-Century New England

Town/Colony	% of 20-Year-Olds Surviving to at Least Age 60
Plymouth Colony	68.0
Andover, Mass.:	
first generation (born 1640–69)	60.4
second generation (born 1670–99)	65.7
Ipswich, Mass	71.7
Middlesex County, Mass	54.9

SOURCES: John Demos, *A Little Commonwealth: Family Life in Plymouth Colony* (New York, 1970); Philip J. Greven, Jr., *Four Generations: Population, Land, and Family in Colonial Andover, Massachusetts* (Ithaca, N.Y., 1970); Susan Norton, "Population Growth in Colonial America: A Study of Ipswich, Mass.," *Population Studies*, XXV (1971), 433–52; Carol Shuchman, "Examining Life Expectancies in Seventeenth-Century Massachusetts" (unpublished paper, Brandeis University, 1976).

censuses, tax lists, meetinghouse plans, and other town records. Once again, a process of linkage to individual death-dates yields an approximate life table. Unfortunately, however, the deaths of some persons were not recorded; in their case, the method substitutes the latest date when they are known to have been living. (This amounts to an assumption that all such persons died in the same year when they last appeared in the records—clearly an overestimate of mortality.) The result is a two-part construction, establishing an upper bound of survivorship (based only on known ages at death) and a lower one (based on the entire population at risk). Presumably the actual rate of survivorship lay somewhere in between.

This method has been applied, for the present study, to two different local communities: Hampton, New Hampshire, in the year 1656, and Wethersfield, Connecticut, in 1668. In the case of Hampton the chances of survival from age twenty to at least sixty were between 61 and 77 percent. Mean expectation of life for young people of about twenty was between 42 and 47 years; for children of ten the comparable figures were 47 and 52. (Life expectation at birth is harder to calculate, owing to irregular recording of infant deaths; however, a reasonable guess would be 44 to 52 years.) The Wethersfield figures are not quite so high. Survivorship from age twenty to age sixty was between 60 and 62 percent. Life expectation at the same age averaged 39 years; at age ten it was 44 years. (See Table 3.)

Taken altogether, these results do not markedly alter the picture of longevity drawn in the earlier studies. Additional research is surely needed, especially for larger, more commercial communities (such as Boston and Newport), but the evidence at hand is already fairly substantial. And all of it suggests that survival to old age was a better-than-even prospect for young people in colonial New England.

Some questions remain, finally, about actual contacts between the generations in these various communities. Granted that the "old" cohort was a definite part of the larger age structure and granted, too, that survival to old age was a reasonable expectation for many of the New Englanders, we may yet wish to know how much, and in what ways, the elderly were known by others in the course of daily experience.

A partial answer to this question can be obtained by reconverting the age-structure percentages to numbers of individual people. If an average New England community contained some 500 inhab-

TABLE 3. Life Expectation and Survivorship in Hampton (1656) and Wethersfield (1668)

	Hampton (Males and Females)			Wethersfield (Males Only)		
	High	Low	Medium	High	Low	Medium
Life expectation in years at ages:						
0–5	51.6	44.0	47.8	60.3	54.0	57.2
6–15	52.5	47.1	49.8	45.9	41.2	43.6
16–25	47.0	41.9	44.5	39.3	38.3	38.8
% surviving to age 60 from ages:						
0–5	57.5	41.0	49.3	63.6	57.1	60.4
6–15	62.9	48.6	55.8	42.9	38.0	40.5
16–25	77.4	60.9	69.2	61.8	60.0	60.9
% surviving to age 70 from ages:						
0–5	40.0	25.3	32.7	50.0	39.3	44.7
6–15	51.4	36.7	44.1	26.2	22.0	24.1
16–25	58.1	43.5	50.8	26.4	22.5	24.5

NOTE—"High" estimates are based exclusively on persons for whom the age at death is known. "Low" estimates add to this group all other known residents of the town; in their case death is assumed to have occurred immediately following the date when they are last noted as being alive in any of the extant records. The latter procedure clearly overstates actual mortality. The "medium" estimate simply averages "high" and "low" in each instance. The age groups for which the material is organized (see left-hand column) are meant to produce rough averages for ages 2, 10, and 20. The findings are summarized accordingly in the text. The Wethersfield data are flawed, for the earliest ages, by incomplete recording of infant deaths. The findings for the age group 0–5 years cannot, therefore, be given much credence.

SOURCES: Town Book of Hampton (New Hampshire), 2 vols. (manuscript holdings, in Town Offices, Hampton, N.H.); Town Votes of Wethersfield, Connecticut (manuscript volume, in Connecticut State Library, Hartford, Conn.); and family reconstitution by the author.

itants, and if the over-sixty cohort was normally in a range of 4 to 7 percent, the total of the elderly in such places must have fallen between a low of about 20 and a high of about 35. The figures for smaller and larger communities are easily computed according to the same principle. This, in short, is a simple way of gauging the numerical possibilities for intergenerational contact.

However, the question should ideally be refined so as to express specific forms of contact. Here one rather direct line of approach suggests itself. New Englanders of the colonial era were deeply responsive to family ties, and perhaps the family offered, at least to some of them, early and powerful experience of older people. The reference, of course, is to grandparents. It was suggested in a scholarly review of some years ago that grandparents may well have been a "New England . . . invention, at least in terms of scale," but we have as yet virtually no published research on the subject. The following paragraphs are intended as a very modest beginning.[16]

What can be learned about the qualitative aspects of the grandparent-grandchild relationship? There is, for a start, important evidence scattered through probate records, in the form of direct bequests by elderly testators to their children's children. Examples abound: "I give and bequeathe unto my granddaughter Hubbard . . . one cow [named] Primrose." And, "As concerning my grandchild Abiel Sadler . . . I do give and bequeathe unto the said Abiel Sadler my last and tools belonging to my trade." And again, "I give and bequeathe unto my five grandchildren, the children of my son John Neal by Mary his now wife: viz. Jeremiah, John, Jonathan, Joseph, and Lydia Neale, fifty pounds sterling between them." Considered overall, grandchildren formed the second most important category of beneficiaries in New England wills (surpassed only by the testators' own children). The records also offer passing glimpses of the same impulse at work while the old yet lived—for example, this note in an estate inventory from 1678: "My mother in her lifetime disposed of her wearing apparel by her particular desire to her granddaughter Hannah Blaney."[17]

Moreover, it is clear that grandparents and grandchildren were sometimes involved in the exchange of personal help and services. When young children were orphaned, grandparents might be called to serve *in loco parentis*. For example (a court order), "John Cheney, sr., of Newbury, was chosen guardian to his grandchild, Abiel Chandler, aged about two years"; and also (a clause from a will) "my mind is to bequeathe my two daughters unto my dear mother-in-law Mrs. Elvin in Great Yarmouth, entreating her and my loving father Mr. Elvin, her husband, to take care of them." Sometimes a testator would make these arrangements in advance, contingent on a subsequent remarriage by his spouse: "in case [my wife] . . . shall marry again, then my will is that if my father William White pleases he shall have full power to take my son John home to himself, and have the sole and whole care of his education and power to dispose and order him." Occasionally such transfers occurred even when both parents were living:

"Philip Fowler the elder, of Ipswich, in the presence of Joseph his son and Martha his wife, and with their full consent, adopted as his own son Philip, the son of the said Joseph and Martha."[18]

But if grandparents often cared for young children, the reverse was also true. Again the probates are a useful source. One Massachusetts resident made a special bequest to his granddaughter "because of her diligent attendance on me." Another noted a prior "covenant, or agreement, betwixt myself and my grandchild," according to which the latter was promised valuable properties in exchange "for his managing my affairs."[19] Occasionally such arrangements drew special attention from the courts. Thus "Capt. Thomas Topping of Bradford requested the Governor and this county court to grant an exemption from public service to his grandson, sent by the youth's father from Long Island to help him in his old age in his domestic affairs and occasions." The court responded affirmatively, citing "defects" in the old man's "sight and hearing . . . so far that he needs constant attendance upon his person and occasions."[20]

One final piece of evidence bearing on these cross-generational ties is a striking admonition in an essay on child rearing by the Reverend John Robinson. "Grandfathers are more affectionate towards their children's children than to their immediates," wrote Robinson, "as seeing themselves further propagated in them, and by their means proceeding on to a further degree of eternity, which all desire naturally, if not in themselves, yet in their posterity. And hence it is that children brought up with their grandfathers or grandmothers seldom do well, but are usually corrupted by their too great indulgence."[21]

Unfortunately, the extant records do not support any firm calculations as to the number of children actually "brought up with their grandfathers or grandmothers," though it cannot have been very large. What can be calculated, at least on a limited scale, is the numerical presence of grandparents and grandchildren within a single community. Once again, the demographic reconstruction of Hampton, New Hampshire, provides a test case. In 1680 the population of Hampton included approximately 290 children and "youth" under the age of nineteen. For some 200 of these there is complete information about the survival (and death) of grandparents. Over 90 percent had at least one grandparent currently alive (and, with only a few possible exceptions, resident in Hampton). For obvious reasons, the pattern was strongest with respect to very young children. Thus no child under five years old was altogether without grandparents; indeed, a clear majority of this age group had three or four grandparents still living. But older children were affected too: in the age group ten to nineteen nearly half had at least two grandparents living. These findings refer only to blood relationships; they must be revised upward when step-grandparents are taken into account. For example, over 80 percent of the entire sample population had at least two living grandparents of either the natural or the step variety. (See Table 4.)

The same data can be used to depict sequences of developmental experience with grandparents. Of course, such experience varied markedly, depending on a child's position in the birth order. A first- or second-born child was likely to know all his grandparents in his earliest years and to have two or three still surviving as he entered his teens. A middle child (third-, fourth-, or fifth-born) would have perhaps one grandparent less at each equivalent stage along the way. And a child born near the lower end of the birth order (below fifth) was lucky to have two living grandparents at the outset and one as he grew up. The median age, for each of these groups, at the death of the last surviving grandparent was approximately twenty-five (first or second born), twenty (third through fifth), and twelve (fifth or below).

These materials suggest, in conclusion, that grandparent-grandchild ties were (potentially) close and (relatively) widespread. Many children received exposure, in the context of family experience, to the ways and wisdom of the elderly. There was much interest and affection in this relationship, at least on the side of the grandparents; occasionally, there was co-residence and mutual dependence. Whether or not grandparenthood was "invented" in early New England, it certainly seems to have flourished there.

Our detour into demography has, at length, yielded valuable results. It is evident now that old age was (1) an attribute of a small but not insignificant portion of local populations in colonial New England; (2) a life stage at which many individual persons would eventually arrive; and (3) a human condition which almost everyone, as children, observed from close up. There was, in short, nothing intrinsically unusual about growing—or being—old.

IV

Just as the culture at large recognized old age as a distinct time of life, so too were elderly people conscious of their own aging. They thought about it and talked about it, and in various ways they

TABLE 4. Children and Grandparents in Hampton, N.H., 1680

Children's Ages	No. of Living Grandparents					No. of Children
	0	1	2	3	4	
	% of Children with Living Grandparents					
0–4	0	18	23	46	9	79
5–9	2	30	38	23	8	53
10–14	6	31	31	25	8	36
15–19	44	34	6	16	0	32
	% of Children with Living Grandparents and Step-Grandparents					
0–4	0	5	22	20	53	79
5–9	2	11	32	17	38	53
10–14	6	6	36	17	36	36
15–19	44	13	22	13	9	32

NOTE: Sample includes 200 children from 61 different families. Another 86 children, from 21 families, were not included because of incomplete information as to the pertinent relationships.

SOURCES: Nathaniel Bouton, ed., *Documents and Records Relating to the Province of New Hampshire* (Manchester, N.H., 1867); Town Book of Hampton (New Hampshire), 2 vols. (manuscript holdings, in Town Offices, Hampton, N.H); Joseph Dow, *History of the Town of Hampton, New Hampshire* (Salem, Mass., 1893); Sybil Noyes, Charles T. Libby, and Walter Goodwin Davis, *Genealogical Dictionary of Maine and New Hampshire* (Baltimore, 1972).

acted from a particular sense of age-appropriate needs and requirements. "Old age is come upon me," wrote one man in beginning a letter to his brother.[22] "This is a matter of great grief to us now in our old age," stated an elderly couple when obliged to testify in court about a quarrel with their daughter and son-in-law.[23] And there were set phrases included in various New England wills as a kind of explanatory preface: "being ancient and weak of body," or "considering my great age and many infirmities accompanying the same," or (more grandly) "having through God's goodness lived in this world unto old age, and now finding my strength to decay [and] not knowing how near my glass is run."[24] A rare seventeenth-century diary throws indirect light on the same point: Thomas Minor of Stonington, Connecticut, began to note his *birthdays* only when he had reached a fairly advanced age. Minor kept his diary on a regular basis from 1653 until 1684, covering the age span in his own life from forty-five to seventy-six. The first of his birthday entries was made on April 23, 1670: "I am 62 years old." The next one came in 1675; similar notations appeared thereafter on April 23 of each year.[25]

But if Thomas Minor counted the passing of his final years, numbers were not ordinarily the chief criterion of old age. As noted earlier, chronological age was imprecisely specified on various official documents (e.g. those that certified release from military training); indeed, there is reason to think that many New Englanders did not know, or did not care, precisely how old they were. More important, surely, to the subjective experience of aging was the dimension of physical change and decline. Again we should note the relevant phrases from the wills, which invariably coupled "age" and "infirmity." (In some cases the former is subsumed under the latter, e.g. "by reason of my great age and other infirmities."[26])

We touch here on the physiology of aging, a particularly difficult and elusive subject for historical study. Lacking any equivalent of modern geriatric data, we can reach only a few relatively simple conclusions. Demographic materials suggest that the elderly in colonial New England were not a great deal more liable to mortal illness and injury than their twentieth-century counterparts. Life expectation at age sixty appears to have been at least 15 years; at seventy, about 10; at eighty, about 5. These figures are only a little lower than the comparible ones for today.[27]

Yet there is no doubting the depth of the association between age and physical depletion in the minds of the New Englanders. "Infirmity," "deformity," "weakness," "natural decays," "ill savors," "the scent of rottenness"—such terms recur throughout their writings on old age. Anne Bradstreet's poem, "The Four Ages of Man," puts some especially pungent description into the mouth of a fictive representative of the elderly:

> My almond-tree [gray] hairs doth flourish now,
> And back, once straight, begins apace to bow,
> My grinders now are few, my sight doth fail,
> My skin is wrinkled, and my cheeks are pale.
> No more rejoice at music's pleasant noise,
> But do awake at the cock's clanging voice.
> I cannot scent savors of pleasant meat,
> Nor savors find in what I drink or eat.
> My hands and arms, once strong, have lost their might.

I cannot labor, nor can I fight.
My comely legs, as nimble as the roe,
Now stiff and numb, can hardly creep or go.[28]

Language such as this implies some particular element of stress and shock in physical aging, as typically experienced by the New Englanders. Was morbidity, their presumed vulnerability to illness, the critical factor here? Perhaps, but the evidence is ambiguous at best. Old people in this setting certainly suffered from frequent and protracted bouts of illness, but so, too, did many others still young and vigorous. Compared with our own time, morbidity was not particularly age-specific; hence this factor alone would not well distinguish old age from the earlier phases of life.

In fact, the literary materials on old age make little reference to illness; they stress instead the loss of capacities and skills. Just here lies an important clue. We can scarcely overestimate the importance of physical exertion in pre-modern times: the "strong arms" and "nimble legs" of which Bradstreet wrote were directly engaged by the work of the farm, by the routines of the household, by travel, transport, and a host of other quite mundane activities. In this context "infirmity" was bound to have a very intense and focused meaning. It seems likely, moreover, that physical decline was often postponed until quite late in life. Even today the pace of such changes varies markedly from one individual and one setting to another. Strength, coordination, and a relatively trim physique can be preserved long after youth has passed by regular exertion and exercise. Conversely, physical decline can be hastened, given a more inactive and sedentary style of life. We may suppose that the former pattern best approximates the experience of the early New Englanders. If so, their subjective sense of the life course was shaped accordingly. Nowadays physical aging is typically, and often powerfully, associated with the passage from youth to middle age; here, indeed, lies an important part of what we call the "mid-life crisis." But centuries ago an equivalent crisis may well have marked the entrance to old age.

These considerations should help to explain some distinctly negative undertones in the attitudes of the elderly toward their own aging. For despite the "honor" that was prescribed as their due, few of them seem to have enjoyed being old. Increase Mather advised "aged servants of the Lord" to "comfort themselves with this consideration: God will never forsake them. They may live to be a burden to themselves and others; their nearest relations may grow weary of them; but then the Everlasting Arm will not grow weary in supporting them." Another minister, himself past sixty, urged his peers to avoid any semblance of "foolish" or "ridiculous" behavior. He particularly warned against "everything in our old age which may look as if we were loath to be thought old" (for example, "vain boasts of the faculties yet potent with us"). The same man three decades earlier had rebuked all "trifling and childish and frolicsome sort of carriage" in the elderly. "We cannot reverence you unless your grave looks, as well as your gray hairs, demand it of us."[29]

It is impossible to know how well and widely the elderly maintained their "grave looks"; there are, however, some grounds for speculation about their "gray hairs." Near the end of the seventeenth century there arose in Massachusetts a sharp controversy about what one man called "the evil fashion and practices of this age, both in apparel and [in] that general disguisement of long, ruffianlike hair."[30] Centrally at issue here was the widespread and increasing use of wigs. The Reverend Nicholas Noyes of Salem sounded the tune for those who opposed the trend in a lengthy "Essay Against Periwigs." "The beauty of old men is the gray head," wrote Noyes, citing Scripture. And he continued:

> The frequent sight of gray hairs is a lecture to men against levity, vanity, and youthful vagaries and lusts . . . Others are obliged to rise up before and honor the old man, the demonstrative token of which is his gray hairs. But strangers to old men cannot so well distinguish of the age they converse with, when youthful hairs are grafted on a gray head, as is oftentimes [true] in the case of periwigs . . . and when periwigged men are known to be old, though they do the utmost to conceal their age, yet such levity and [vanity] appears in their affecting youthful shows as renders them contemptible and is in itself ridiculous.[31]

Unless the concerns of the Reverend Noyes were totally removed from social reality, some elderly men preferred to appear younger than they actually were. In fact, although the symbolic importance of gray hair was everywhere recognized in New England society, the context of such recognition was at least occasionally pejorative. Here is a small but revealing instance. Two older, and locally eminent, men—Mr. Edward Woodman and Capt. William Gerrish—were arguing opposite positions before the town meeting at Newbury, Massachusetts. In the heat of the debate Gerrish made a slighting reference to the "gray hairs" of his adversary. The rejoinder was reported as follows. "Mr. Woodman said . . . that his gray hairs would stand were Capt. Gerrish, his bald pate, would." It appears, then, that the "demonstrative token" of old age did not always elicit "honor," even from the elderly themselves.[32]

But, most of all, old people dreaded what the young might think or say about them. "Our patience will be tried," declared Cotton Mather in his mid-sixties, "by the contempt which the base may cast upon us, and our beholding or fancying ourselves to be *lamps despised* among those who see we are going out." A generation earlier Increase Mather had composed a tract of *Solemn Advice to Young Men*, in which the following idea was central: "It is from pride that young men do not show that respect to their superiors, or unto aged ones, which God commandeth them to do. . . . Such especially whose parts and abilities are through age decayed: proud youth despiseth them." And elsewhere the same author wrote: "To deride aged persons because of those natural infirmities which age has brought upon them is a great sin. It may be they are become weak and childish: They that laugh at them on that account, perhaps if they should live to their age will be as childish as they. And would they be willing to be made a laughing-stock by those that are younger than they?" This preoccupation with ridicule, whatever its relation to actual experience, implies a considerable insecurity in the aged themselves.[33]

There is one additional device—an ingenious creation of Professor Fischer—for measuring attitudes toward aging. For all sorts of reasons and in many different settings, individual persons misreport their age. This tendency is evident even in our own day, but it was much stronger in pre-modern times. Its chief manifestation in colonial New England was in rounding off the exact figures to one or another multiple of ten: 29 sometimes became 20, 62 might be reduced to 60, and so forth. When these reports are grouped (for example, in a large population listing), they show an effect now called "age heaping"—a disproportionate clustering around the aforementioned ten-year levels. Thus five times as many people were likely to report themselves as being 50 than 49 or 51, and likewise at the comparable points across the entire age spectrum. An interesting question in the present context is the direction of this effect: did more people round off up or down? Did they prefer to make themselves a little older, or a little younger, than they

TABLE 5. Age-Heaping Ratios for Essex County, Mass.

Age Levels	Ratios	Age Levels	Ratios
28–29	.920	51–52	.460
30	2.349	58–59	.627
31–32	.667	60	3.222
38–39	.754	61–62	.535
40	3.109	68–69	.511
41–42	.621	70	2.764
48–49	.976	71–72	.748
50	3.232		

NOTE: Ratio is computed by dividing actual number of persons of given age by number of those reporting the age. ("Actual" numbers are determined by averaging reports for age levels within five years on either side of the given age.) Under-reporting can be expected within two years on either side of the 10-year intervals, and the material has been organized accordingly. All data have been taken from age reports given by deponents in the Quarterly Courts of Essex County, Massachusetts, 1636–1682.

SOURCE: *Records and Files of the Quarterly Courts of Essex County, Massachusetts*, 8 vols. (Salem, Mass., 1911–21).

actually were? If the former was true, we may infer an "age bias," if the latter, a "youth bias," in the population at large.[34]

The only large sample of age reports currently available for seventeenth-century New England involves 4,000 Massachusetts residents called as witnesses in trial proceedings before the Essex County Court. (Witnesses were normally required to give their age before testifying.) The critical data are the ages within two years on either side of the "rounded" levels (e.g. 38, 39, 41, and 42, around the figure 40). Careful tabulation shows a preponderance of *down*-ward revision—in short, a "youth bias." Curiously, the pattern weakens somewhat near the end of the life cycle and disappears altogether for people of seventy or more. (However, the sample numbers are probably too small at the most advanced age levels to yield results of much significance.) Taken as a whole, the material reveals a personal orientation toward aging that was markedly unfavorable. (See Table 5.)

V

We turn now from subjective considerations—the "self" view of aging—to questions of social experience. Here, fortunately, the evidence becomes more ample and varied, and conclusions can be more directly reached.

There is the important matter of domestic location—how the elderly were positioned in relation to home and family. In our own time a considerable number of people lose their homes or find themselves living alone as they pass into old age; the "nest" is empty or abandoned altogether. In colonial New England such conditions were rare.

For one thing most New Englanders continued to live in their own homes even after their children had grown up and moved out to begin separate families. There was no general pattern of relocation for the elderly, in response to altered needs for space and/or altered resources. In fact, the issue of resources was crucial: the majority of older people were well off when compared with those in other age groups. The evidence for this conclusion derives from tax and probate records. When age is correlated with wealth—both for decedents whose estates were inventoried and for household heads "rated" by local tax committees—a consistent pattern emerges. Average wealth was lowest for the age group of men in their twenties, rose strongly through the thirties and forties, reached a peak in the fifties, and declined gradually thereafter. The reason for the eventual downward turn is obvious: men past sixty were deeding away property to their grown children. But if the elderly had somewhat less wealth overall then their "middle-aged" neighbors, they also had greatly lessened needs. A household in which the head was forty or fifty would normally include a number of young children, including some too young to earn their keep by contributing meaningful labor. The same household twenty years later was much reduced in size. The (by then) old man might well have no one to support beyond himself and his spouse. He might perhaps be labor-poor, but much of his property would remain intact—and this with many fewer mouths to feed.

Let us try to visualize other aspects of the same man's situation. His children, grown by now and established in their own households, were still a part of his social environment. The lure of new lands and fresh opportunities might possibly have drawn one or two to distant locations, but most of them would be living nearby.[35] The details of these intrafamilial relationships remain obscure, and there are no grounds for assuming any special elements of closeness or harmony. Still, the simple fact of the children's presence within the same community was important in its own right.

To the children, moreover, were added the grandchildren. We have already considered this matter from the standpoint of the young, but we should look at it again from the farther end of the life course. Most elderly New Englanders were grandparents many times over. Two cases, drawn from the Hampton data, will serve to illustrate the common pattern. Isaac Perkins and his wife Susannah were born in about 1610 and 1615, respectively, and were married in 1636. Together they produced twelve children, three of whom died before reaching adulthood. Of the nine children who married, eight produced children of their own. The eldest of the latter was born in the year 1660, when Isaac and Susannah were approximately fifty and forty-five. (Susannah bore her last child the year after the birth of her first grandchild—thus did the generations overlap.) By 1670, when Isaac had reached sixty and Susannah fifty-five, there were 12 grandchildren; a decade later there were 31. When Isaac died in his mid-seventies there were 39 grandchildren; when Susannah died at eighty-four the total had risen to 56. By this time, too, great-grandchildren were starting to arrive. The experience of the Perkins's neighbors, Morris and Sarah Hobbs, can be more quickly summarized. The first of their grandchildren was born when they were about fifty and forty-five years old; 13 more had arrived by the time Sarah died at sixty-one. Morris lived on another twenty years. At age seventy he had 23 grandchildren; at seventy-five, 36 grandchildren and 2 great-grandchildren; and at eighty-six (the year of his death), 44 grandchildren and 17 great-grandchildren.[36]

These figures seem extraordinary by the standards of our own day; but, given the arithmetic of high fertility and surprisingly long life expectation, they are plausible—indeed, virtually inevitable. Again, they establish only the statistical presence of grandchildren, but stray notations gleaned from court records hint at the qualitative dimensions as well. Thus one old man remembered dispatching his grandson "to the cowhouse . . . to scare the fowls from my hogs' meat"; another, needing spectacles to read a neighbor's will, "sent his grandchild Mary . . . to Henry Brown's and she brought a pair."[37]

While presumably enjoying their proximity to kin, elderly couples preferred to look after themselves. Thus Governor George Willys of Connecticut wrote, in concluding a long letter to his son about various matters of inheritance: "I will say no more; for you know, as it is a way of prudence, so it is also my judgment and shall be my practice, not so to dispose to any child but that (God preserving my estate in an ordinary course of Providence) I may have to maintain myself and not to be expecting from any of them." An Essex County farmer invoked the same prinicple when asked whether he might not increase the "portion" of his married son: "He answered that he had been advised to keep his estate in his own hands as long as he lived, and as they were young and lusty, they could work to get themselves necessaries."[38]

It appears that most elderly couples succeeded in remaining self-sufficient; as noted previously, they retained on average a relatively high level of wealth. Inevitably, however, there were some circumstances in which they required assistance. The most obvious and certainly the most common of these was widowhood. When a man died, his wife was placed in a position of some doubt, or even of jeopardy. If she was still young, she might look forward to remarriage, which would automatically supply her deficiencies. But if she was elderly, her prospects that way were greatly reduced.[39] In any case, her rights to her late husband's property must be secured through appropriate action in the courts. The principle of the widow's "thirds" was long established in custom and in common law: she would have the use of one third of the family lands during her lifetime, plus full title to a third of all movable properties. Often the wills of the decedents added explicit provisions for her daily maintenance. "My will is," wrote Jonathan Platt of Rowley, Massachusetts, shortly before his death, "that my two sons John and Jonathan do provide well for my beloved wife, and that they let her want nothing that is needful for herself so long as she remaineth my widow." Similarly, John Cheney of Newbury "enjoined" his son Daniel to supply his widow with "whatever necessaries . . . her age shall require during the time of her normal life."[40]

In many instances these arrangements were spelled out with extraordinary precision. Typically, the widow was guaranteed appropriate space for her lodging ("the parlor end of the house . . . with the cellar that hath lock and key to it"); access to other parts of the household ("free liberty to bake, brew, and wash, etc., in the kitchen"); a fuel supply ("firewood, ready cut for the fire, at her door"); furnishings ("the bedstead we lie on, and the bedding . . . thereunto belonging . . . and the best green rug . . . the best low chair . . . and a good cushion"); and household implements ("pots, kettles, and other vessels commonly made use of"). She might also be given domestic animals ("two cows, by name Reddy and Cherry, and one yearling heifer") and regular assistance in

caring for them ("kept for her use by my heir, wintered and
summered at his charge, and brought into the yard daily, as his
own is, to be milked"); a share in the fruits of the garden ("apples,
pears, and plums for her use"); and a means of travel ("a gentle
horse or mare to ride to meeting or any other occasion she may
have"). Less often—and chiefly where there was considerable
wealth in the family—she would receive regular payments in
money or produce ("eight pounds per year, either in wheat, barley,
or indian corn"); personal service ("Maria, the little Negro girl,
to be with her so long as my wife lives"); and special sustenance
for mind ("the book called *The Soul's Preparation for Christ*, and
that of Perkins upon the creed") or body ("her beer, as she hath
now").[41]

These provisions, reflecting a broad range of probate settle-
ments, directly involved the testator's children (and heirs) in his
widow's care. Frequently, the widow herself was empowered to see
to their fulfillment. Thus one man directed that "if Nathaniel [his
son] fail of anything he is to do for my wife, my will is that he shall
forfeit ten pounds every year he fails"; and another left all his
movable property for his wife to bequeathe "to my children accordingly as she shall see cause and they deserve, in their carriage
and care of her in her widow's estate."[42]

Sometimes elderly men made formal arrangements for their
own care, following the death of their wives. One old farmer
promised a bequest to his son-in-law on condition that the latter
"be helpful towards the maintainance of him while he lived." And
a second, noting his "weak body . . . and solitary condition,"
conveyed his entire estate to a relative who would "find and provide for me wholesome and sufficient food and raiment, lodging,
attendance, washing, and other necessaries, as well in sickness and
weakness of old age as in health." Similar arrangements could be
made even while both spouses survived: "I bequeathe to my cousin
Daniel Gott all my neat cattle and sheep and horse-carts, chains,
plow, and tools . . . in consideration that he is to remove his family
and come to live with me and my wife at Lynn during our lives
and carry on our husbandry affairs."[43]

The care of old and infirm persons did not, of course, always
require a legal document; sometimes it was managed informally,
or simply developed out of familial closeness and affection. Here
and there the wills afford a restrospective glimpse. "Forasmuch as
my eldest son and his family hath in my extreme old age and
weakness been tender and careful of me," wrote one testator, in

identifying the chief beneficiary of his estate. And another—a woman making "her last will . . . upon her death bed"—particularly remembered "my daughter Ann, in consideration of her staying with me in my old age and being helpful to me."[44]

When such "considerations" were not specifically acknowledged in probate documents, they might give rise to legal proceedings. The relatives, even the children, of the deceased were quite ready to put a price on their "tendance and care." When the courts settled the estate of Jeffrey Massey of Salem, Massachusetts, a son filed bills for his "charges . . . with my father and mother in the time of their age and weakness . . . both food, physick, and tendance for the space of four years." Occasionally, such claims were disputed by other family members, and witnesses were called to establish particulars. Thus one dutiful son was described as having taken such "great trouble and care . . . with his mother that he could hardly spare time to go abroad about his business." A second had helped his aging father to frame a house, and—a witness remembered—"did almost all the work"; the father had subsequently remarked "that his son was the best friend he had."[45]

At least a few elderly New Englanders, finding themselves in need of help but without relatives living nearby, were obliged to appeal to public authority. Usually, the local meeting would supervise their care. Particular tasks and services were assigned to individual townspeople, who were subsequently reimbursed out of town funds. The financial accounts of Watertown, Massachusetts, for the year 1670 reveal the process at work in a specific case:

	£ s d
To widow Barlett, for dieting old Bright	10-00-00
To John Brisko, for a bushel of meal to old Bright, and 7 loads of wood	01-11-06
To William Perry, for working for old Bright	00-03-03
To Michael Baristow . . . for cloth for a coat for old Bright	01-18-06

The recipient of these charities, Goodman Henry Bright, was "old" indeed—ninety-five years old by the best available estimate. The local records of New Haven, Connecticut, show a similar pattern of involvement with a man identified only as "old Bunnill." The town voted him "maintainance" of two shillings per week (among other benefits). Eventually "the town was informed that old Bunnill is desirous to go to old England . . . where he saith he hath some friends to take care of him," and public funds were authorized to pay for his passage.[46]

VI

The problems created by infirmity were only part of the experience of old age. Many elderly New Englanders retained a substantial capacity for work, for public service, for ordinary forms of social intercourse. The same thing is true (even more so) of our own society; however, the context is now vastly changed.

To make comparisons with modern conditions is to spotlight at once the issue of "retirement." Did the New Englanders retire from active pursuits as they crossed the "borders" of old age? A quick answer is likely to be negative. Almost any set of local records can be used to observe old people in the postures of their workaday world. Here are several examples, culled from the files of the Essex County Courts: Edward Guppy, age sixty, "employed by Edmund Batter to mow salt-water grass in the marsh"; William Nichols, age seventy, hauling grist to the local mill; Evan Morris, age sixty-six, working "as a retainer" at the Rowley ironworks; William Boynton, age sixty-eight, hired by a merchant to transport "rugs and blanketing" to Boston; George Kesar, age sixty-seven, tanning leather for his neighbors; James Brown, age seventy-four, still active as a glazier; Edmund Pickard, age sixty, "master of the [ship] Hopewell"; Wayborough Gatchell, age seventy, continuing "her services as midwife";—and most remarkable of all—Henry Stich, age one hundred and two, working as a "collier" at Saugus.[47]

Older people also played substantial parts in public service. The various governors and assistants of Plymouth Colony seem, for the most part, to have retained their offices until the end of their lives.[48] The careers of New England clergymen were also extremely prolonged; in a sample of thirty-five only three were terminated by causes other than death.[49] Individual cases are no less impressive. In 1700 the Hampton vital records gave special notice to the death of "Henry Green, Esq., aged about 80 years, for several years a member of the Council of New Hampshire until by age he laid down that place, but a Justice till he died which was the 5th of August." In 1669 Thomas Minor of Connecticut made the following entry in his diary: "I am by my own accounts sixty-one years old. I was by the town this year chosen to be a selectman, the town's treasurer, the town's recorder, the brander of horses, [and]

by the General Court recorded the head officer of the train-band
. . . one of the four that have the charge of the militia of the whole
county, and chosen and sworn commissioner and one to assist in
keeping the county court."[50]

Yet the total picture is complicated, and the evidence does not
run entirely in a single direction. Henry Green, after all, had "laid
down" one of his offices on account of age. And there are similar
indications in the careers of other public officials, such as the
renowned Samuel Sewall of Massachusetts. When Sewall was
sixty-five years old and long established as a magistrate, he put his
name forward for the position of chief justice. The Governor was
evasive in his reply: he recognized Sewall's qualifications, but "did
not know but that by reason of my age I had rather stay at home."
Sewall "humbly thanked His Excellency" for this mark of personal consideration—and continued his quest for the position.
Ten years later, though, it was Sewall himself who cited the
exigencies of age: "I went to the Lieut. Governor and desired to lay
down my place in the Superior Court. I was not capable to do the
work, and therefore was not willing to hold the place." Thomas
Munson of New Haven made a similar plea while seeking, at the
age of sixty-three, to retire from his post as lieutenant in the local
militia: "He said he had been an officer to the company long, and
had willingly served to the best of his ability, but he finds such
decays in himself, and thereby [feels himself] unfit to serve in that
place and office any longer."[51]

Lieutenant Munson seems, in fact, a better case for our purposes than either Green or Sewall; as a local (not provincial)
official he more closely approximated average experience. But the
study of local leadership is most usefully pursued by way of collective biography. Were selectmen, for example, often men of advanced years? Might they continue to serve until the time of their
death? A modest sample of Hampton selectmen has been investigated with these questions in mind, and the results establish distinct trends in relation to age. The largest portion of the sample
comprised men between the ages of fifty and fifty-nine years (32%);
the next largest group were in their forties (27%). A smaller though
still considerable number were in their sixties (18%); however, very
few (3%) were seventy or more. (See Table 6.)

A subsample drawn from the same material yields an even
clearer picture of withdrawal from officeholding in relation to old
age. The nineteen selectmen in this group all served at least three
terms, and all died at ages of at least seventy. Their mean age at the

TABLE 6. Selectmen's Ages, Hampton, 1645–1720

Age Group	Number of Selectmen	Percentage of Sample
20–29	2	1
30–39	31	19
40–49	45	27
50–59	52	32
60–69	30	18
70+	5	3

SOURCES: Town Book of Hampton (New Hampshire), 2 vols. (manuscript holdings, in Town Offices, Hampton, N.H.); Joseph Dow, *History of the Town of Hampton, New Hampshire* (Salem, Mass., 1893); Sybil Noyes, Charles T. Libby, and Walter Goodwin Davis, *Genealogical Dictionary of Maine and New Hampshire* (Baltimore, 1972).

time of final service was 65.8. Only one of them actually died in office, and fully 90 percent lived at least five years longer. Indeed, the average interval between final term and time of death was more than ten years. (See Table 7.)

Thomas Minor qualifies as a "local leader" in another settlement (Stonington, Connecticut), and his diary offers a unique opportunity to study the career of an older man in detail. Minor's varied responsibilites as an officeholder in his sixty-second year have already been noted. He continued to serve, in similar ways, for some while longer, and was even "employed in the country's service about the Indian War" at the age of sixty-seven. After seventy, however, the pattern seems to have changed. The diary makes no further reference to officeholding, or indeed to any other form of public responsibility.

Minor's life as a private citizen is also fully chronicled; and again there is evidence of vigorous activity well into his seventh decade. Here is the record of a typical month in this sixty-sixth year.

The third month is May and hath 31 days. Friday the first: this week I made my cart. Friday the 8: a town meeting. Monday the 11: I and my wife was at New London. Friday the 15: we pulled down the chimneys. Monday the 18 day: Thomas Park began to build. Friday 22: we shore the sheep and had home one of the manteltrees. 25: we had a town meeting. 27 Wednesday: we laid out Bay grants on the east side [of the] Poquatuck.

Beyond the age of seventy Minor continued to work on his farm, to visit friends, and the like—but at a somewhat reduced pace. By this time he was receiving considerable help from others. In the spring of his seventy-third year, for example, one of his sons "took all the corn to sow and to plant, to halves for this year."[52]

One final test has been made of the effects of aging on Thomas Minor by coding the events noted in his diary on a simple "self/ other" basis. Minor's reports of his own activity (any event in which he is himself a chief agent) can then be compared with his references to the doings of others. The results, briefly summarized,

TABLE 7. Selectmen's Ages at Retirement and Death, Hampton, 1645–1720

Age in Last Term as Selectman		Interval between Last Term and Death	
Age	No.	Years	No.
59 or less	2	0	1
60	1	1	...
61	1	2	1
62	...	3	...
63	2	4	...
64	2	5	1
65	...	6	...
66	1	7	3
67	...	8	1
68	4	9	1
69	1	10	2
70	...	11	1
71	1	12	1
72	3	13	1
73	...	14	1
74	1	15	2
75 and over	...	16	...
		17	3
		18 and over	...

NOTE: The sample includes 19 men, all of whom died at ages of at least 70, having served at least three terms as town selectman. The average age of selectmen at retirement was 65.8 years. The average interval between last term and death was 10.3 years.

SOURCES: Town Book of Hampton (New Hampshire), 2 vols. (manuscript holdings, in Town Offices, Hampton, N.H.); Joseph Dow, *History of the Town of Hampton, New Hampshire* (Salem, Mass., 1893); Sybil Noyes, Charles T. Libby, and Walter Goodwin Davis, *Genealogical Dictionary of Maine and New Hampshire* (Baltimore, 1972).

show a modest lowering of the ratio (self/other) as Minor passed into his sixties and a very marked decline in his early seventies. Though this measure is admittedly crude, it allows us to glimpse the deeper rhythms of aging in one particular case. (See Table 8.)

The retirement question evokes, in sum, a complex and divided answer from the New England source materials. It appears that most men past sixty voluntarily reduced their activities in work and/or public service. Yet nearly always this was a gradual process—a quantitative rather than a qualitative change—and rarely did it lead to complete withdrawal. There were two main categories of exceptions: those ministers and magistrates whose exalted rank exempted them from even a partial retirement, and those among the ordinary folk whose "infirmities" were simply incapacitating.

We should understand, too, that the idea of retirement was not entirely unknown to the New Englanders. The Mathers, father and son, both remarked on its sometimes painful aspect. "It is a very undesirable thing for a man to outlive his work," wrote Increase, "although if God will have it so, His Holy Will must be humbly and patiently submitted unto." Cotton pursued the same thought at greater length: "*Old folks* often can't endure to be judged less able than ever they were for *public appearances*, or to be put out of *offices*. But good sir, be so wise as to *disappear* of your own accord,

TABLE 8. Thomas Minor's Activities Pattern

Years	Minor's Age	Self	Other	Ratio
1654–57	45–48	98	25	3.92
1658–61	49–52	149	44	3.39
1662–65	53–56	129	50	2.58
1666–69	57–60	151	52	2.90
1670–73	61–64	156	88	1.77
1674–77	65–68	144	68	2.12
1678–80	69–71	88	41	2.15
1681–84	72–75	88	93	.95

NOTE: The first six calendar months of each year (January-June) were coded for references to personal activity. The numbers in the "self" column include all activities in which Minor describes himself as a major participant. The "other" column includes activities which Minor attributes to other people in his local environment. (The following kinds of activity are *not* included: marriages, births, deaths, sicknesses, injuries, holidays, and religious observances.)

SOURCE: Thomas Minor, *The Diary of Thomas Minor, Stonington, Connecticut*, ed by Sidney Miner and George D. Stanton (New London, Conn., 1899).

as soon and as far as you lawfully may. Be glad of a *dismission* from any *post*, that would be called for your *activities* . . . Let your *quietus* gratify you. Be pleased with the *retirement* which you are dismissed unto." Comments like these seem to reach forward to our own time; yet the similarities, no less than the differences, are liable to overstatement. The germ of modern retirement was present three centuries ago—but only the germ. What was then gradual, partial, and indefinite at many points has now become abrupt, total, and rigid in its specific applications. A *process* with intrinsic biological connections has become a *moment* plucked from the calendar.[53]

VII

There is one more aspect of social experience to which our study should make such approach. We established at the outset that the normative code of colonial New England was decidedly favorable to old age. "Honor," "respect," even "veneration" were the terms most frequently used in prescribing attitudes toward the elderly. Younger people were urged to "rise up . . . before the old man," and conduct themselves with "a bashful and modest reverence." The elderly, for their part, would "use a kind of authority and confidence in their words and carriage." But were these precepts actually followed in practice?[54]

Occasionally—very occasionally—some fragment of the documentary record allows us to glimpse people of different generations dealing with one another face-to-face. A few of these have been noted already: one more will be added now. Two residents of Scituate, Massachusetts, were arguing on a summer day in 1685 about a debt for several bolts of cloth. One was Nathaniel Parker, age twenty-three, the other Edward Jenkins, age approximately sixty-five. (The site was Jenkins's own house.) Tempers flared— and then: "Nathaniel Parker ran to Edward Jenkins and took Edward Jenkins by the collar or neckcloth that was about Edward Jenkins's neck; and Nathaniel Parker said. 'God damn me, if thou were not an old man, I would bat thy teeth down thy throat'." Shall we count this as an expression of deference to age? Literally, yes—but, in context, no. Certainly it was not the kind of deference prescribed in the published literature on old age. Parker's comment seems, in fact, to imply *contempt* for his adversary's weakness as an "old man."[55]

The connotations, in the New England setting, of the word

"old" deserve our further consideration. Again and again in local records we find elderly people mentioned in a special way: "old Bright," "old Bunnill," "old Woodward," "old Hammond." Their given names are, in effect, discounted, and age itself becomes an identifying mark. (For younger persons the pattern of reference was consistently otherwise, with given name and surname appearing together.) This usage was not, moreover, a matter of indifference to the elderly themselves. Increase Mather made the point very clearly: "To treat aged persons with disrespectful or disdainful language only because of their age is a very criminal offense in the sight of God; yet how common is it to call this or the other person 'old such an one,' in a way of contempt on the account of their age." A list of householders in the records of Watertown, Massachusetts includes five men designated in precisely this way. Their age, as determined from independent evidence, ranged between sixty-seven and eighty-two. Interestingly, there are other men of equivalent age on the same list who are not called "old." The distinction was one of social rank, pure and simple. Thus the man listed as "Simon Stone" (age 72) was wealthy, a deacon of the Watertown church, and frequently a town officer, whereas "old Knapp" (age 74) was a sometime carpenter of little means and no public responsibilities whatsover; the implicitly pejorative prefix could only be applied in the latter instance. Probate documents afford parallel evidence on the same matter. Testators called themselves "ancient" or "aged" (when they referred to such things at all), but never "old." It was fine to describe property that way—"old housing," "old lumber," "old cows," or whatever—but not the person writing the will.[56]

So much for individual terms of reference; were there no formal—perhaps even institutional—expressions of deference to the elderly? One possibility comes immediately to mind. The inhabitants of New England towns, like people in traditional communities everywhere, came together at regular intervals to honor in a ceremonial way their deepest values and spiritual commitments. Weekly (sometimes twice weekly) they gathered for worship in the village meetinghouse. On these occasions they reaffirmed not only the shared basis of their corporate life but also the hierarchical arrangement of its constituent parts. Every meetinghouse was carefully "seated"—that is, all adult members of the community occupied places assigned to them in accordance with their individual status. The basic principle was: the higher the rank of the person involved, the closer his (or her) seat to the front. The official criteria for making these status evaluations invariably included "age"—along with "estate" (i.e. wealth), "office" (public service), "dignity of descent" (pedigree), and "pious disposition." Here, then, was an unmistakable mark of the preferment deemed appropriate for older citizens.[57]

However, we may well ask to what extent, and in what ways, this criterion of status was actually applied in conjunction with all the others. Fortunately, detailed seating plans have survived in various town archives; unfortunately, they have not as yet received much systematic attention from scholars.[58] That they deserve such attention seems indisputable, for they are virtual "sociograms" that would tell us much about seventeeth-century life. But for the moment we have only one set of results on which to base some very tentative conclusions.

The Hampton (New Hampshire) meetinghouse was seated in the early months of the year 1650. The design of the building was relatively simple. The pulpit was on the north side; just below, and roughly in the middle of the floor space, was a "table" where men of the highest rank were privileged to sit. The other male members of the congregation were assigned to one of eight benches in the west and southwest parts of the building; the women occupied comparable places directly opposite. We do not know precisely what instructions were given to the committee that took the measure of every single individual named on the plan; nor could we, in any case, deal with such intangibles as "pious disposition." We can, however, investigate the importance of age and wealth relative to one another.[59]

Predictably, age and wealth were substantially intercorrelated overall, and in many individual cases they simply cannot be distinguished. Older men (or women) who were also wealthy had a double claim on front-row seats, while those who were both young and poor invariably occupied places near the rear. In other instances, however, there were sharp status discrepancies—for example, a young and rich man, or an old and poor one. What places were thought appropriate for *them*? Most such people were seated on the basis of wealth. Thus, William Cole, at seventy-nine the oldest resident of the town, had one of the lowest tax assessments; he sat on a back bench in the meetinghouse. By contrast, Thomas Ward, only thirty-one but already near the top in terms of wealth, was assigned to the front row. Cole and Ward belonged to a group of eleven particularly discrepant cases (i.e. where rankings for age and wealth are at least three quintiles apart). In nine of these wealth was the decisive factor for seating position; in the remaining two there was something of a compromise. An additional group of sixteen cases, in which the age/wealth discrepancy was less (two quintiles), includes eight which favored wealth, four which favored age, and four in the compromise category. (See Table 9.) In truth, the Hampton materials from 1650 are not the best imaginable for this inquiry. The town has been settled barely a decade earlier, and few of its people were truly old. The data strongly suggest that age in general was not an important criterion of social rank, but it might yet be shown, on other evidence, that *old* age was treated specially.

VIII

This long excursion through an extremely varied and tortuous body of source materials is now at an end; there remains, however, the question of central themes and tendencies. Here we may allow ourselves to be quite openly speculative. Moreover, we may look for help to the social sciences, where the study of aging has been more extensively pursued.

We are told by sociologists that the status of old people in any culture turns on a cluster of institutional factors. Among these the following seem especially important: (1) property ownership; (2) the possession of strategic knowledge; (3) the predominant modes and styles of economic productivity; (4) an ethos of mutual dependence (or, conversely, of "individualism"); (5) the importance of received traditions (especially religious ones); (6) the strength of family and kinship ties; and (7) the range and character of community life.[60]

Measured against this checklist, the position of the elderly in colonial New England looks strong. We know, for example, that some of them (merchants, artisans, or particularly successful yeomen) controlled large amounts of property, and that old people in general were well off in comparison with most other age groups. "Strategic knowledge" was also an acquirement of the elderly. Farming, marketing, domestic craftsmanship: these things they knew at least as well as their younger neighbors. In addition, and more important, they controlled a variety of significant information about the community's past. Much that pertained to the

TABLE 9. Hampton Meetinghouse Plan, 1650,
Status-Discrepant Cases (in quintiles)

Name	Age	Wealth	Seat	
			Row	Section
3-4 quintiles:				
Brown, John	4	1*	1	s
Cole, Eunice	1	5*	5	e
Cole, William	1	5*	3	w
Elkins, Gershom	2	5*	3	w
Elkins, Mary	2	5*	4	e
Fuller, Francis	4	1*	1	s
Huggins, John	2	5*	5	w
Moulton, Margaret	5	3	2	s
Sanborn, John	5	2	2	s
Ward, Margaret	5	1*	2	s
Ward, Thomas	4	1*	1	w
2 quintiles:				
Brown, Sarah	3	1*	1	e
Estow, Mary	1	3*	2	s
Estow, William	1*	3	Table	...
Green, Henry	5	3	2*	w
Leavitt, Thomas	4	2	2	w
Marston, Mary	5*	3	3	s
Moulton, William	4	2	2	s
Philbrick, Thomas	1*	3	1	s
Sanborn, Mary	4	2	2	e
Sanborn, Mary	5	3	2	e
Sanborn, William	5	3*	2	s
Sleeper, Thomas	3	5*	3	w
Smith, Deborah	3	5*	4	e
Smith, John	3	5*	4	w
Swain, William	1*	3	1	s
Taylor, Anthony	2	4*	3	s

NOTE: Ages have been computed (in some cases estimated) from vital records, genealogies, etc. Wealth was determined by averaging positions (relative to all taxpayers) on tax lists of 1647 and 1653.

*Indicates that this factor was apparently given greatest weight in seating assignment.

SOURCES: Town Book of Hampton (New Hampshire), 2 vols. (manuscript holdings, in Town Offices, Hampton, N.H.); Joseph Dow, *History of the Town of Hampton, New Hampshire* (Salem, Mass., 1893); Sybil Noyes, Charles T. Libby, and Walter Goodwin Davis, *Genealogical Dictionary of Maine and New Hampshire* (Baltimore, 1972).

settlement of legal questions—boundaries, contracts, the details of ownership—was never written down and had to be recalled in some appropriate forum by those who could bear personal witness from long ago:[61] "John Emery, sr., aged about eight-one years, testified that about forty years ago he saw laid out to William Estow, then of Newbury, a four acre lot . . ."[62] In a society only partially literate and without comprehensive record keeping, the *memories* of old people gave them a certain advantage.

Other institutional factors can be followed in the same way, and to roughly parallel conclusions. Thus, the fact that early New England was land-rich and labor-poor enhanced the productive value even of marginal workers (such as the "aged and infirm"). The principle of reciprocity was established at the very core of the value structure: "we must be knit together in this work as one man," John Winthrop had said in a famous speech prepared enroute to the New World.[63] The force of tradition was appreciated, even venerated, throughout New England society. Most people were well supplied with kinfolk, and there was a vigorous network of neighborly relationships. In sum, the position of the elderly was supported, even enhanced, by prevailing social arrangements. Certainly their power and influence compare very favorably with what obtains for their counterparts in our own time.

We may now feel that we have finally uncovered the basis in social reality for the dictum "honour old age." And yet, too, we have seen how that dictum was subject to varying interpretations and was often directly controverted by actual behavior. To understand these somewhat paradoxical findings it is necessary to look beyond social structure to considerations of psychological functioning. Just here there are valuable suggestions to be taken from anthropological research on the aging process in a variety of premodern cultures.[64] This research can be summarized only at the risk of gross oversimplification, but the effort is worth making nonetheless. In most, if not all, premodern settings, the elderly occupy a position of far greater social importance than is true in our culture; in part, this is based on their control of valuable resources and in part on their presumed status as being "closer to God." But however powerful, they are not invariably secure. In some societies the elderly elicit great respect and affection; in others they are the object of deep resentment and mistrust and live in a chronic state of fear. The difference is not based on their institutional position, which may be strong in both cases, but rather on the predominant style of affective and interpersonal life in the culture—in technical language, on a culture-specific capacity for "object relations."

In fact, older people everywhere are liable to be considered alien, different, strange; given certain preconditions, they arouse in their younger culture-mates mixed feelings of awe and apprehension. In a society where interpersonal relations are more or less relaxed, where there is little subjective tension between the claims of self and others, where psychosocial conditions favor the formation of "internalized objects"—in such a society the elderly remain secure. As one scholar has written, "by keeping his 'object' status the older person avoids becoming the *stranger* . . . [who arouses] fear and revulsion."[65] Or, to put the matter in still another way, the older person is experienced fully as an individual being, in whom the past (what he formally was) and the present (what he now is) are implicitly joined. But things are not always so. In other societies, where object relations are narrower, less differentiated, more narcissistic in tone, the aged (and sometimes also the very young) are distinctly at risk. Their strangeness is highlighted and often deeply feared. No matter what their socioeconomic power and offical prestige, they are vulnerable to various forms of covert, even overt, attack.

These two situations are, in fact, the opposite ends of a single spectrum. And we must now ask where on that spectrum to locate the New Englanders. The question seems impossibly large, yet the materials discussed in the preceding pages suggest at least some parts of an answer. The people of early New England had many strong and admirable qualities, but there was indeed something problematic about their "object relations." In the doings of many of them one feels a thin edge of psychic vulnerability—a sense of self somewhat insecurely held, a view of others not fully three-dimensional. Committed always to goals of "peaceableness," they often disappointed themselves; conflict, inner and outer, was the actual condition of their lives. The doubt, the distress, the occasional rages which fueled such conflict are manifest all through the documentary record they have left to us.[66] Inevitably, under such conditions, there was some narrowing of their perception and understanding of others. The very qualities of "otherness" were hard for them to appreciate. Their view of the American Indians, for example, was notoriously constricted: Indians must behave in all things like Englishmen, else they are "savages" and "beasts."[67] Even their attitudes toward their children expressed a certain lack of empathy. They were determined, insofar as possible, to "beat

down" infantile expressions of willfulness, and they insisted on
confronting their young with painful reminders of sin and death.[68]
This does not imply an absence of parental love but simply an
inability to credit fully the inherent childishness of childhood.

And what about the bearing of such considerations on old age?
First, the elderly themselves were burdened with an especially difficult experience of their own aging. Growing old always creates
some narcissistic imbalance, but for persons who are already sensitive on that count the problems are greatly compounded. Here,
then, is one way to account for the "peevish," "suspicious," and
"complaining" character usually attributed to the aged in colonial
New England. But we must also look at the matter in terms of
what the others, the not old, contributed. For reasons related to
their own character structure, they were frequently unable to "see"
the elderly in a way that embraced the full richness of human
individuality; to them the old person was indeed something of a
stranger.[69] Thus they tended to stereotype him (calling him "old
such an one"), to fear him (especially for his alleged "covetous-
ness"), and indirectly to ridicule him (witness the figure of the
"staggering, spewing, old drunkard" and the metaphor of the tree
with its leaves off).

To summarize these rather diverse and cross-cutting materials,
we may say that the position of the elderly in early New England
was sociologically advantageous but psychologically disadvanta-
geous. Their control of important resources seemed to command
honor and respect, but not affection or sympathetic understanding. Simone de Beauvoir has written that the only sure protection
for old people is "that which their children's love provides."[70] And
precisely here the situation of old New Englanders was doubtful.

IX

We have pressed about as far as our evidence will carry us—
perhaps, indeed, a bit farther. And yet there are important questions which we have scarcely touched. As noted at the start, most of
the available data concerns aging in men, but sooner or later we
will need to find some parallel way of investigating the experience
of women. Another major *lacuna* in the present treatment is the
matter of attitudes toward death. In all cultures and epochs the
elderly must anticipate death by one means or another, and such
anticipation was expressed with great emphasis in early New England. But death makes a huge subject in itself; fortunately, there

are other studies which approach it more directly and at considerable length.[71]

One further issue should be confronted, if only in a speculative way, before we conclude. Can we say anything about our material in relation to historical change? Were the central tendencies in aging gradually altered as time and circumstance moved them along? And, seen in retrospect, what was their eventual direction?

There is reason to think that the position of elderly people was *improved* between the middle of the seventeenth and the middle of the eighteenth century. Some of our own measures, when applied to later materials, strongly suggest as much. "Age heaping," for example, begins to show an age bias after 1700. A meetinghouse plan from the year 1774 seems to give strong priority to age (as compared with wealth). Wigs were increasingly designed to make their owners look older, not younger, than was actually the case— likewise the sartorial fashions of the eighteenth century. The unfriendly undertones, so persistent in seventeenth century literature on old age, appear to have faded thereafter.[72] It seems possible, then, that the decades immediately preceding the American Revolution were a time of maximum advantage for old people.

If so, we may well wonder *why*, and it is worth recording certain fragments of an explanation suggested by the present research. We have learned that life expectation for the early New Englanders was surprisingly long. Evidently, they survived to old age in numbers unequaled elsewhere in the colonies or in old England across the seas.[73] Perhaps, under these conditions, the aura of strangeness around elderly people was gradually dissipated.

Perhaps, too, there were complementary changes of inner life. This possibility is hard to explore on an empirical basis, but it does link up with well-known themes in New England historiography. Thus the harsh lines of Calvinist belief are thought to have softened somewhat after the middle of the seventeenth century. The balance of social concern tipped away from religion toward secular experience; in terms of cultural types, the shift went "from Puritan to Yankee."[74] There was, moreover, an ecological shift spanning roughly the same time period. The "wilderness," in which personal security and cultural integrity both seemed at risk, gave way to a settled society with its various protections and amenities.[75] New England character was modified accordingly. The claims of self were now more freely acknowledged, and this, in turn, broadened the psychic space available for experiencing others—indeed, "otherness" in general.[76] The inherited social core remained intact (Anglo-Saxon, Christian, adult, and effectively male), but people on the margins were less vulnerable to implicit or explicit stereotyping. And old people were particular beneficiaries.

A further consideration—no easier to specify, but probably no less important—was the meaning of age in a society still relatively new. The example of the "planters" seems to have gained deeper and deeper significance as the decades passed. These redoubtable men and women were widely remembered as the roots of the growing community. History had opened to them a uniquely creative path, and they had followed it unswervingly; their success in "settlement" would remain, for all their descendants, an achievement of stunning proportions. To be sure, the effect on their immediate descendants was problematic: the sons of the planters found it hard to measure up. Tension built to a peak in roughly the third quarter of the seventeenth century; the religious controversies of that era represented, in part, a crisis in age relations.[77] The members of the settler generation were leaving "the earthly stage" enroute to their reward beyond. Their deaths occasioned special comment in local diaries, and some notably elaborate funerals; perhaps, too, there was a connection with the dominant religious motif of "declension." There were signs of sharpened age consciousness overall: increasingly, for example, individual people identified themselves as being old.[78]

In time, of course, the "crisis" eased. Gone as a living presence, the planters survived as the heroic figures of legend. The ambivalence they had aroused in their own children yielded to the unqualified admiration of succeeding generations. The residue of this process was a growing regard for age: old people were closest to the hallowed beginnings of New England, and that alone gave them a certain cachet. Here is Samuel Sewall, writing in his diary in the spring of 1726: "The honored, ancient, elder Faunce . . . kindly visited me. *Laus Deo.*" Thomas Faunce was then about 80, and had served for many years as deacon of the first church of Plymouth. He is said to have "kept in cherished remembrance the first settlers, many of whom he well knew. He used to identify the rock on which they landed."[79] Another venerable link to the same era, Goodwife Ann Pollard, was memorialized in a famous portrait of 1721. She was then past one hundred, old enough to have come to Boston with the very earliest settlers. Indeed she was known as a raconteur of that experience, and claimed to have been the first person ashore—a spry girl of ten, leading Governor Winthrop and his colleagues onto the site of what would later become New England's greatest city.[80]

When Ann Pollard was young, Charles I was king of England and Sir Walter Raleigh had only just died: by her last years Benjamin Franklin was already a young man. The life of the "honored, ancient, elder Faunce" ran from the old age of William Bradford to the childhood of John Adams and Thomas Jefferson. Franklin, Adams, and Jefferson would, in time, lead a political revolution and launch a new nation on its collective life course. Franklin was the most famous old man of the Revolutionary era, and indeed he capitalized on that fact. Adams and Jefferson, too, would be admired, even "venerated," in their old age. And yet a new "revolution in age relations," as Professor Fischer has called it, was coming: "young America" of the nineteenth century was less and less inclined to acknowledge the claims of age. From the ambivalent circumstances of the settlement period—the main concern of the present essay—old people's experience had moved through a long cycle of change. More cycles, more changes would follow. It was, and is, a fascinating story, which historians are only beginning to tell.

NOTES

1. David Hackett Fischer, *Growing Old in America* (New York, 1977).

2. Cotton Mather, *Addresses to Old Men and Young Men and Little Children* (Boston, 1690), 1.

3. Increase Mather, *Two Discourses* (Boston, 1716), 120; *Records of the Colony and Plantation of New Haven from 1638–1649*, Charles J. Hoadly, ed. (Hartford, Conn., 1857), 375; *Records of the Colony or Jurisdiction of New Haven, From May, 1653 to the Union*, Charles J. Hoadly, ed. (Hartford, Conn., 1858), 602.

4. *Records and Files of the Quarterly Courts of Essex County, Massachusetts*, 8 vols. (Salem, Mass., 1911–21), I: 336, 179, 187, 380.

5. John Robinson, "Of Youth and Old Age," in *The Works of John Robinson*, Robert Ashton, ed., 2 vols. (Boston, 1851), I: 253; Anne Bradstreet, "The Four Ages of Man," in *The Works of Anne Bradstreet*, Jeannine Hensley, ed. (Cambridge, Mass., 1967), 61–62.

6. Cotton Mather, *Addresses to Old Men and Young Men and Little Children*, 6; Increase Mather, *Two Discourses*, 65; Nicholas Noyes, "An Essay Against Periwigs," in *Remarkable Providences*, John Demos, ed. (New York, 1972), 215. On the "mediative" position of the elderly, see David Gutmann, "The Cross-Cultural Perspective: Notes Toward a Comparative Psychology of Aging," in

Handbook of the Psychology of Aging, James E. Birren and K. Warner Schaie, eds. (New York, 1977), 302–26.

7. William Bridge, *A Word to the Aged* (Boston, 1679), 3–4; Cotton Mather, *Addresses to Old Men and Young Men and Little Children,* 37–38.

8. *Ibid.,* 40.

9. *Ibid.,* 37.

10. Bridge, *A Word to the Aged,* 11.

11. *Records and Files of the Quarterly Courts of Essex County, Massachusetts,* VII: 156–57.

12. Linda Auwers Bissell, "Family, Friends, and Neighbors: Social Interaction in Seventeenth-Century Windsor, Connecticut" (unpublished Ph.D. dissertation, Brandeis University, 1973), 40.

13. *Documents and Records Relating to the Province of New Hampshire,* Nathaniel Bouton, ed., 2 vols. (Manchester, N.H., 1867), I: 424.

14. "Census of Bristol in Plymouth Colony, Now in Rhode Island, 1689," in *New England Historical and Genealogical Register, XXXIV* (1880), 404–5; John Demos, "Families in Colonial Bristol, R.I.: An Exercise in Historical Demography," in *William and Mary Quarterly,* 3rd ser., *XXV* (1968), 40–57.

15. Philip J. Greven, Jr., *Four Generations: Population, Land, and Family in Colonial Andover, Massachusetts* (Ithaca, N.Y., 1970); Susan Norton, "Population Growth in Colonial America: A Study of Ipswich Mass.," *Population Studies, XXV* (1971), 433–52; John Demos, *A Little Commonwealth: Family Life in Plymouth Colony* (New York, 1970); Carol Schuchman, "Examining Life Expectancies in Seventeeth-Century Massachusetts" (unpublished paper, Brandeis University, 1976).

16. John M. Murrin, "Review Essay," *History and Theory, XXI* (1972), 238.

17. *The Probate Records of Essex County, Massachusetts,* 3 vols. (1916–20), *II*: 61, 54, 50.

18. *Ibid.,* I: 150, 76; II: 108, I: 132.

19. *Ibid.,* III: 266, 11.

20. County Court Records, New Haven County (manuscript volume, Connecticut State Library, Hartford, Conn.), leaf 153.

21. Robinson, "Of Youth and Old Age," I: 246.

22. "Correspondence of Thomas Leeds and William Leeds," in *Remarkable Providences, 1600–1760,* Demos, ed., 151.

23. *Records and Files of the Quarterly Courts of Essex County, Massachusetts,* IV: 81.

24. *The Probate Records of Essex County, Massachusetts,* III: 13, 268, 141.

25. Thomas Minor, *The Diary of Thomas Minor, Stonington, Connecticut, 1653 to 1684.* Sidney Miner and George D. Stanton, eds. (New London, Conn., 1899), 95, 128, 135, 141, 148, 154, 160, 166, 172, 183.

26. *The Probate Records of Essex County, Massachusetts,* II: 441.

27. Demos, *A Little Commonwealth,* 192; Shuchman, "Examining Life Expectancies in Seventeenth-Century Massachusetts," 15.

28. Bradstreet, *The Works of Anne Bradstreet,* 62–63.

29. Increase Mather, *Two Discourses,* 105; Cotton Mather, *A Brief Essay on the Glory of Aged Piety* (Boston, 1726), 27; Cotton Mather, *Addresses to Old Men and Young Men and Little Children,* 37.

30. *The Probate Records of Essex County, Massachusetts,* I: 332.

31. Noyes, "An Essay Against Periwigs," in *Remarkable Providences,* Demos, ed., 215.

32. *Records and Files of the Quarterly Courts of Essex County, Massachusetts,* IV: 123.

33. Increase Mather, *Two Discourses,* 99.

34. Fischer, *Growing Old in America;* Ansley J. Coale and Melvin Zelnick, *New Estimates of Fertility and Population in the United States* (Princeton, N.J., 1963).

35. Greven, *Four Generations;* Bissell, "Family, Friends, and Neighbors"; Kenneth A. Lockridge, "The Population of Dedham, Massachusetts, 1636–1736," *Economic History Review XIX* (1966), 318–44.

36. Family reconstitution by the author.

37. *Records and Files of the Quarterly Courts of Essex County, Massachusetts* V: 356; IV: 343.

38. The Willys Papers. *Collections of the Connecticut Historical Society, XXI* (1924), 71; *Records and Files of the Quarterly Courts of Essex County, Massachusetts,* VII: 269.

39. Rates of remarriage have been calculated for seventeeth-century Windsor, Connecticut. Among younger widows (those whose first marriage lasted less than 10 years) the portion remarrying was 67%; among older ones (with a first marriage lasting at least 20 years) the comparable figure was under 25%. (Bissell, "Family, Friends, and Neighbors," 56.) The actual number of widows in local populations at given points in time can be calculated from census, land, and tax records. In 1670 the Connecticut towns of Hartford, Wethersfield, and Windsor were found to contain a total of 335 heads of household; 18 of those (slightly more than 5%) were widows. A land-allotment list from New Haven in 1680 contained the names of 220 proprietors, 25 of whom were widows (11%). (The Willys Papers, 191–99; *New Haven Town Records, 1662–1684,* Franklin Bowditch Dexter, ed. [New Haven, Conn., 1919], 405–10.) For an excellent discussion of widowhood in a slightly later period, see Alexander Keyssar, "Widowhood in Eighteenth-Century Massachusetts: A Problem in the History of the Family," *Perspectives in American History,* VIII (1974): 83–119.

40. Richard B. Morris, *Studies in the History of American Law* (New York, 1930), ch. 3; Demos, *A Little Commonwealth,* 85; *The Probate Records of Essex County, Massachusetts,* III: 390; II: 52.

41. *Ibid.,* III: 223; *The Mayflower Descendant* (Journal of the Massachusetts Society of Mayflower Descendants), XXX: 101; *The Probate Records of Essex County, Massachusetts,* II: 221, 63; *The Mayflower Descendant,* II, 184–85; *The Probate Records of Essex County, Massachusetts,* I: 96, II: 346, 263; *The Mayflower Descendant,* II: 184–85; *The Probate Records of Essex County, Massachusetts,* III: 48; *The Mayflower Descendant,* II, 184–85; *The Probate Records of Essex County, Massachusetts,* II: 64; III: 175.

42. *Ibid.,* II: 239; *The Mayflower Descendant,* XVII: 34–36.

43. *Ibid.,* XI, 92–93; *The Probate Records of Essex County, Massachusetts,* III: 357; *ibid.,* II: 251.

44. *The Mayflower Descendant,* VI: 81; *The Probate Records of Essex County, Massachusetts,* II: 171.

45. *The Probate Records of Essex County, Massachusetts,* III: 150–51; *ibid.,* 377; *ibid.,* II: 144.

46. *Watertown Town Records*, 5 vols. (Watertown, Mass., 1894–1911), I, part one, 101; *New Haven Town Records, 1649–1662*, Franklin Bowditch Dexter, ed. (New Haven, Conn. 1917), 116, 208.

47. *Records and Files of the Quarterly Courts of Essex County, Massachusetts*, III: 276; V: 29; V; 396, V: 188; VI: 77; VI: 44; VII: 326; VIII: 345; II: 97.

48. Demos, *A Little Commonwealth*, 174–75.

49. Fischer, *Growing Old in America*, 45.

50. Joseph Dow, *The History of the Town of Hampton, New Hampshire*, 2 vols. (Salem, Mass., 1893), II: 740; Minor, *The Diary of Thomas Minor*, 207–8.

51. Samuel Sewall, *The Diary of Samuel Sewall*, M. Halsey Thomas, ed., 2 volumes (New York, 1973), II: 884, 1051; *New Haven Town Records, 1662–1684*, Dexter, ed., 331.

52. Minor, *The Diary of Thomas Minor*, 133, 122–23, 165.

53. See William Graebner, *A History of Retirement* (New Haven, Conn., 1980).

54. Noyes, "An Essay Against Periwigs," in *Remarkable Providences*, Demos, ed., 215; Robinson, "Of Youth and Old Age," I: 251.

55. The Davis Scrapbooks (manuscript volumes, Pilgrim Hall, Plymouth, Mass), II: 8.

56. Increase Mather, *Two Discourses*, 98–99; *Watertown Town Records*, I, part one, 53; *The Probate Records of Essex County, Massachusetts*, II: 4, 229, 315, 345, 352, 426, 441; III: 13, 183, 187, 278, 329, 375.

57. For a general discussion of the principles and practices involved in seating the meetinghouse, see Robert J. Dinkin, "Provincial Massachusetts: A Deferential or a Democratic Society," (Ph.D. dissertation, Columbia University, 1968). For specific instances, see *Watertown Town Records*, I, part one, 47; and Town Votes of Wethersfield, Connecticut (manuscript volume, Connecticut State Library, Hartford, Conn.), leaves 115–16.

58. In addition to the material in Dinkin, "Provincial Massachusetts: A Deferential or a Democratic Society" (see fn. 57), meetinghouse plans are discussed in Ola Winslow, *Meetinghouse Hill 1630–1783* (New York, 1952) and John Coolidge, "Hingham Builds a Meetinghouse," *New England Quarterly, XXXIV* (1961): 435–61. Coolidge makes some approach to the issue of seating and social rank in seventeenth-century Hingham, Massachusetts and, clearly, believes that older people received favored positions. However, he does not seem to have attempted systematic comparisons of the age factor with other variables such as wealth.

59. Town Book of Hampton, two volumes (manuscript volumes in Town Offices, Hampton, N.H.), I, leaves 28–29.

60. See Irving Rosow, *Socialization to Old Age* (Berkeley, Calif., 1974).

61. The effect of this pattern can be measured quantitatively from legal records. Persons over 60, we know, comprised 10% or so of the total adult population of early New England; yet they account for some 13% of all witnesses in court cases in seventeenth-century Massachusetts. (*Records and Files of the Quarterly Courts of Essex County, Massachusetts*, I-VIII, *passim.*

62. *Records and Files of the Quarterly Courts of Essex County, Massachusetts*, VII: 194.

63. John Winthrop, "A Model of Christian Charity," in *Puritan Political Ideas*, Edmund S. Morgan, ed. (New York, 1965), 92.

64. See Gutmann, "The Cross-Cultural Perspective: Notes Toward a Comparative Psychology of Aging,"

65. *Ibid.*, 315-16.

66. See, for example, *A Little Commonwealth*, 136-38; Richard L. Bushman, *From Puritan to Yankee: Character and the Social Order in Connecticut, 1690-1765* (Cambridge, Mass., 1967), 20-21; Darrett B. Rutmen, "The Mirror of Puritan Authority," in *Puritanism and the American Experience*, Michael McGiffert, ed. (Reading, Mass., 1969), 65-79; Emory Elliot, *Power and the Pulpit in Puritan New England* (Princeton, N.J., 1975), 76-80.

67. See Neal Salisbury, *Manitou and Providence: Indians, Europeans, and the Making of New England, 1500-1643* (New York, 1982).

68. See Demos, *A Little Commonwealth*, 136-38; Edmund S. Morgan, *The Puritan Family* (New York, 1966), ch. 3; David Stannard, "Death and the Puritan Child," in *Death in America*, David Stannard, ed. (Philadelphia, 1975), 9-29.

69. The innerlife roots of this pattern are perhaps best understood by way of the theory known (in psychoanalytic circles) as the "psychology of the self." See, in particular, the writings of Heinz Kohut: e.g. *The Analysis of the Self* (New York, 1970) and *The Restoration of the Self* (New York, 1977).

70. Gutmann, "The Cross-Cultural Perspective: Notes Toward a Comparative Psychology of Aging," 314.

71. *Ibid.*

72. *Ibid.*, 315-16.

73. See, for example, David Stannard, "Death and Dying in Puritan New England," *American Historical Review*, LXXVII (1973), 1305-30; Stannard, *The Puritan Way of Death: A Study of Religion, Culture and Social Change* (New York, 1977).

74. See, for example, Bushman, *From Puritan to Yankee*; Kenneth A. Lockridge, *A New England Town: The First Hundred Years* (New York, 1970).

75. See William Cronon, *Changes in the Land: Indians, Colonists, and the Ecology of New England* (New York, 1983).

76. See Richard D. Brown, *Modernization: The Transformation of American Life* (New York, 1976), ch. 5; John Demos "Introduction," in *Remarkable Providences*, Demos, ed., 19-22.

77. See Murrin, "Review Essay," 235-40.

78. The history of religious "declension" has been one of the grand themes of early New England historical studies. See especially the writings of Perry Miller: e.g. his *The New England Mind: From Colony to Province* (Cambridge, Mass., 1953), 3-118.

79. Sewall, *The Diary of Samuel Sewall*, Thomas, ed., II: 1046. The comment associating Faunce with "the first settlers" (and with Plymouth Rock) is found in a note accompanying an earlier edition of Sewall's diary. See *Collections of the Massachusetts Historical Society*, fifth series, VII: 376.

80. James T. Flexner, *First Flowers of Our Wilderness* (Boston, 1947), 46-49.

CHAPTER VIII

History and the Formation of Social Policy Toward Children: *A Case Study*

Among all my efforts of professional outreach, none was so keenly felt as my service with the Carnegie Council on Children. The cause seemed transcendantly right, the prospects hopeful, the process in every sense engaging. In addition, my fellow Council-members made a group as lively and stimulating as any I have ever known. I realized at the start that I had much to gain from the Council myself—much to learn, to enjoy, to wonder about. Nothing else in my experience would bring me to such close quarters with "practitioners" of many kinds. Community organizers in rural Mississippi; health-workers in the slums of Mexico City; advocates for native American rights; nutritionists, day-care teachers, lawyers, social workers, labor leaders, corporation executives; not to mention a Vice President, a Cabinet secretary, plus assorted United States Senators and Congressmen: thus the roster of our working contacts. For a humble historian it was heady stuff.

Yet, also from the start, two questions nagged at me, and they nag at me still. The first concerned my own participation—my contribution to the work of the Council—what I might give in

This essay was first published in David J. Rothman and Stanton Wheeler, eds., *Social History and Social Policy* (New York, 1981), 301–24. (Copyright © 1984 by Academic Press, Inc.)

*return for all I gained. I was scarcely a practitioner in the above-
mentioned sense; indeed, the methods of such practice, and even its
language, were sometimes incomprehensible to me. I knew things
about children—children who had lived and died centuries ago—
but . . . how to connect? The essay that follows records my somewhat inconclusive struggle to find a way. Would I advise a new
Carnegie Council to take a historian on board? My answer is
cautiously affirmative (though I'd like to talk with him/her first).
Would my erstwhile colleagues agree? I'm really not sure.*

*The other question was larger and altogether more important.
Could our Council fully respond to its charge, and make a difference in the lives of American children? Could any such group—
privately organized and funded, and lacking official channels of
contact to the sources of public policy—be finally effective? The
answer depended not just on our own intelligence and good will,
but also on the readiness of others to listen. And, in truth, there
was more apparent readiness when we started (1972) than when we
finished (1977). The early 1970s were a time of gathering hope and
energy around "family" issues (day care, flexible work schedules,
nutritional adequacy for children, pre-school education, and the
like). But by the late '70s, and even more the '80s, much of that
spirit had gone. For reasons beyond our control, our timing was
wrong. Perhaps, too, our findings and recommendations seemed
unwelcome, given their somewhat "radical" tone.*

*Seen in hindsight, the Carnegie Council may perhaps be de-
scribed as a noble failure. But there will surely be more such efforts
in the future—and there should be. Can anyone dispute the need
for "long views" of social reality, for bringing ethics alongside
policy, for efforts of collective consciousness-raising? Results will
inevitably vary, from one instance to the next, but a society like
ours must never stop trying.*

I

In the summer of 1972 one of the nation's leading private founda-
tions, the Carnegie Corporation of New York, laid plans for a
substantial new project in social policy investigation. The focus
was to be America's children and especially its *young* children
(defined as all those under 10 years old). The end in view was a
general reassessment of current situation and future prospects. The
means to this end was the establishment of a study group or

"council"—the Carnegie Council on Children—with a core membership of about a dozen persons.

The procedural format would follow a pattern applied in many previous instances. The council would convene at intervals of several weeks, for perhaps 2 or 3 days at a time, over a span of 3 years. (Eventually, the council would extend its lifetime to nearly 5 years.) Meanwhile, a regular staff, embracing a similar number of people, would conduct full-time investigation at the behest of the council. The chairman of the council was also the director of the staff; otherwise, there was no overlap in the membership of these two bodies.

The charge of the Carnegie Corporation to the council was extraordinarily broad. The corporation had long been involved in promoting research on children's learning—their "cognitive development"—but now something more was involved. As the president of the corporation would later put it, "Increasingly we found ourselves asking whether the ability of children to learn was not linked to many other facets of child development, and whether child development itself was not heavily influenced by its social context."[1] There was a feeling, too, that the country was on the verge of major policy decisions affecting the lives of the young. The recently concluded political struggle over the comprehensive child-care bill of 1971 had produced a negative result—a presidential veto—but new efforts along similar lines seemed likely in the future. Further ahead lay possibilities such as national health insurance, welfare reform, revision of the tax laws, in all of which the interests of children were profoundly concerned. Policymakers at various levels, and in both the public and the private sectors, might thus make good use of a reasoned overview of children's needs and circumstances. Or so it then appeared.

Reflecting the breadth of its assignment, the council and its staff comprised an unusual mix of professional and personal backgrounds. The members included only one certified specialist in child development; for the rest, there were lawyers, teachers, social workers, economists, anthropologists, a pediatrician—all broadly committed to the welfare of children and some with direct experience in one or another field of child advocacy.[2] There were also two historians—one on the council, the other a member of the regular staff.

This essay constitutes, in effect, an offering of retrospective testimony by the council's resident historian. It strives toward three somewhat divergent aims. An opening section summarizes the history *of* the project, to the point in the autumn of 1977 when the council's "core report" was officially published. A middle section explores the place of history *in* the project—that is, the use, non-use, and misuse of historical ideas by the council and its staff. The final section is more openly subjective, posing as it does some pointed questions about the role of a historian *on* the project. A short conclusion attempts to draw a few morals and lessons for the future.

Considered as a whole, the chapter is a literal "case study." It presents, therefore, no general viewpoint, and it needs testing against other, roughly similar cases. That such cases will multiply in the coming years seems likely, given current signs and trends. The architects of projects in one or another area of social policy seem increasingly inclined to incorporate historical "perspectives." Usually this takes the form of including one or more professional historians in staff positions and building historical study into project agendas. Since the pattern—perhaps we should call it simply a tendency—is so new, there is as yet little clarity about substantive goals and strategies. Indeed, in some projects the inclusion of history, and of historians, may prove to be no more than a form of tokenism.

However, it is worth noticing at the outset that discussions of public policy often refer to history without specific plan or intent to do so. One example may be worth at least brief consideration, by way of preface to the main business of this chapter. Shortly before the Carnegie Council came into being, many of the issues it would confront were spotlighted in the public debate over the child-care legislation of 1971. Even a quick review of this debate shows that both sides—supporters and opponents—sporadically cited the impact of historical trends on children's lives. The supporters repeatedly emphasized a sequence of social changes, which made their proposals seem desirable—even necessary. The growing participation of mothers of young children in the labor force and the increased numbers of single parents, for example, virtually required a commitment to the establishment of day-care centers. Meanwhile, opponents of the legislation detected a gradual erosion through time of traditional institutions and values. In their view, day-care centers in particular, and the "child development" viewpoint in general, threatened further damage and decay. Above all, the family stood in jeopardy—and they viewed the family as a prime source of whatever strengths and virtues Americans yet retained. There was in their pronouncements a note of alarm, almost of outrage, lest government action with respect to children should breach a sacred boundary.[3]

It seems promising, from the viewpoint of a historian, to find that history did indeed figure in this debate. But a second look at the material elicits doubts and questions. For one thing, both sides advanced historical ideas in a careless and self-serving way. The "conservatives" (those against the bill) invoked a highly idealized image of the family in the past, with which to contrast its modern "decay." The liberals cherished some sentimental notions of their own. One advocate of day care, writing in the *New York Times*, pinned her case to an alleged change in family structure from "extended" (formerly) to "nuclear" (nowadays):

> Once upon a time [the family] consisted of grandparents, uncles and aunts, cousins, and even friendly neighbors. . . . If we are to preserve the family's strengths for our children, then we must provide mothers with some of the help they used to receive from the extended family. A good day-care program acts as an extended family.[4]

It is not just that such views are inaccurate when measured against what scholars have learned about the history of the family; there is also a sense in which they converge. Both sides of the child-care debate agreed that American life had changed—in some respects for the worse—so far as families are concerned. Their disagreements involved the matter of *response* to change. Liberals proposed new policies to meet the inescapable reality of new circumstances, new pressures, new needs. Conservatives urged redoubled efforts to consolidate what was left of traditional family life. But the *historical baseline* in each case was similar, if not identical. Perhaps, too, it was in some sense irrelevant.

One more thought suggests itself from the welter of claims and assertions in this debate. Public discussion of childhood and the family seems, whenever it turns toward the past, to take on the roseate hues of nostalgia. Somehow this subject, these concerns, are invested with a collective charge of sentiment. The outcome hardly differs from what is often true of individuals, as they recall, with exaggerated fondness, some aspect of their personal pasts.

II

The central business of early council meetings was, necessarily, to define the task. Strategic alternatives were faced, and judgments were made, that would shape the project irrevocably. Several of these bear careful consideration from the standpoint of the present discussion.

The first important judgment came, in effect, ready-made from the officers of the Carnegie Corporation. The council's charge was to "think broadly," to construct insofar as possible a "framework" within which policy programs might be organized, and to explore the "ecology of childhood." But, of course, this left much open territory. Actually, its chief import was to suggest what need *not* be done. The council would not assemble another "shopping list" of specific needs and problems. It would not produce detailed recommendations to deal with one or another situation affecting children. It would not undertake fresh research in child development, family sociology, community organization, or any comparable field. Instead, the project would make use of information already available and evaluate programs already in existence, in the course of fashioning larger perspectives.

But what was the appropriate starting point for this investigation: How even to *begin* reaching toward such large and lofty horizons? In fact, three different approaches were considered at length in early council meetings. The first entailed sustained effort to identify major and continuing trends in American life insofar as they affected children. The second was frankly visionary: One could try to define an "ideal world" for children and then devise plans and principles for making the real approximate the ideal. The third involved a focused examination of a single "problem area," not for the sake of developing programmatic "solutions," but rather as a way of unraveling the complicated web of institutions, values, and scientific information bearing on *every* aspect of children's lives.

At one time or another, the council would make use of each of these strategies. The "problem-area" approach was tried very early, around the specific focus of children's nutrition. It was possible in this way to consider corporate industry (the food companies), advertising, scientific research (on various aspects of nutrition), social contexts (the "family meal"), and even some highly philosophical issues (such as food in relation to mind-body dualism).[5] Significantly, however, the council chose to abandon its nutrition studies within about six months. Many words were committed to paper, but none to print. (Indeed the paper has remained undisturbed in assorted office files ever since.) In retrospect, this appears to have been something of a warm-up for the council—an opening exercise that served to test various ways of working together and to flag certain issues and questions that would reemerge in other guises during the years ahead.

The notion of an "ideal world" did not figure largely in the council's substantive discussions. Perhaps it might have done so in a different setting and with different people, but the project as originally constituted was loaded with "real world" experience. Still, from time to time the council did seek to be "visionary," to glimpse something radically better than the existing order of things. Evidently, this expressed a felt need in all council members to reaffirm the seriousness of the task and to recapture an immediate sense of moral commitment.

The strategy of exploring cultural trends and tendencies proved, in the end, quite central to the council's work. Repeatedly, discussion moved out from the present and immediate experiences of childhood to the temporal flow in which all such experiences are carried along. This point will bear further consideration below.

A second strategic judgment was of another sort. The council was asked to consider children, but where, so to speak, would it find them? Again, various possibilities suggested themselves. There was, for example, the possibility of a direct approach, an effort to hold children alone in view, trusting that help and support might somehow reach them directly. An obvious alternative was to target the child in the family, on the assumption that families are (and will continue to be) the most important setting for early experience. Still another possibility was to set a broad stage and direct attention to whole communities, if it seemed that the welfare of children was actually inseparable from that of other groups (the elderly, the poor, the racially oppressed—not to mention the affluent, the powerful, and those in the "prime of life"). Here, too, no absolute choice seemed prudent, or even possible; however, the strongest emphasis would fall on the *second* alternative. The Carnegie Council on Children became, secondarily, the Carnegie Council on Families. Parents, the council decided, are usually the best "experts" about their particular children; therefore, helping children inevitably means helping families.

A third question presented another difficult choice. Should the council concern itself with *all* American children, or should it concentrate on those who are most immediately and heavily at risk? Was it foolish, perhaps immoral, to link the children of backcountry hollows or inner-city ghettoes with the "comfortable" children of middle-class suburbs? In the end, the council compromised. Some problems affect all American children and families today; to this extent, it seemed important to "think national." But other problems chiefly involve specific groups: The victims of race prejudice and poverty, for example, disproportionately include children. The council's surveys, reports, and recommendations would therefore take aim in both directions.

These decisions on basic issues of strategy evolved rather gradually through the first 6–10 months of council work. During the same period, council members and staff initiated a variety of substantive investigations. As noted already, considerable effort was directed to a study of children's nutrition, and all staff persons were drawn into this to some degree. The staff included no single specialist in this field, so papers and discussions were built from reviews of the published literature and interviews with recognized experts. This would, in fact, become standard procedure for other aspects of the project as well. The staff, as originally constituted, was composed largely of able and resourceful "generalists." Most staff members had broad competence in one or another field (e.g., political science, history, health care, law), but not more specific forms of expertise. Consistent with the original conception of the project, they were expected to gather, sift, and blend together recent research, wherever it might bear on council aims.

In essentially the same spirit, early council meetings were planned as much to gather information from outside sources as to pool ideas from within. The meeting sites were notably varied: LaJolla, California; Davenport, Iowa; Jackson, Mississippi; Little Rock, Arkansas; Denver, Colorado; Madison, Wisconsin—among others. (One meeting was held outside the country, in Cuernavaca, Mexico, in order to foster international comparisons.) On each occasion, there was considerable effort to consult opinion in the immediate locale. In part, this was "expert" opinion (academic and/or professional), but in part, too, it involved lay persons from all parts of the community. The council sought, most especially, to hear from people who were active in advocacy groups at the local level.

After the completion of its first year, the project underwent a partial, and gradual, change of orientation. The early sorting phase was over, and the balance between internal discussion and external opinion-gathering altered somewhat. The latter would never be entirely given up, but more and more the council concentrated on developing its own viewpoint, its own priorities, and its own style. Meeting arrangements reflected this change, in terms both of sites (New York, Boston, and Chicago were increasingly preferred) and of schedules (less time for "field visits" to day-care centers, health units, and the like). There was a complementary trend with respect to staffing. Certain of the generalists who had at the outset occupied central positions now left the project, and their replacements tended to be people of sharper professional definition.

Meanwhile, a sense of overall direction was gradually emerging in the discussions of the council itself. Time after time, consideration of some particular problem area led off to issues and themes of the largest sort. Periodically, council members sought to pull themselves back to the more immediate concerns of children (and families). Yet, the very persistence of this tendency seemed to express a truth of its own—or, if not a truth, at least a conviction in which most members shared. The deficits and the difficulties that most deeply affected children could not be addressed in piecemeal fashion, and the search for their sources led to the very center of the social and economic "system." What was called for, in the long run, involved nothing less than "structural change." (The political persuasion of most council members at the outset could reasonably be described as center-to-liberal, but the project seemed to move them, albeit rather quietly, toward much more radical positions.)

Reflecting this structural orientation, the council staked out some broad areas of study. Questions of economics, most especially the distribution of jobs and income, assumed an absolutely central importance. The impact of technology, of mass society, and of racial and class distinction followed close behind. A provisional agenda of books and reports emerged accordingly. The council would issue a "core report" to summarize its major findings and recommendations. Two additional publications (written by particular staff members "for the council" and expressing council views) would treat at length the issues of inequality in American society and of socialization to American ways and values. Finally, three "background studies" would deal with more specialized problems and themes: handicapped children, minority children in the schools, and the implications for policy of recent research in child development. The actual completion of this agenda was a long and complicated process, and the details have little significance in the present context. However, it is necessary to say something of the outcome, if only to trace the perimeters through which historical ideas did, or did not, finally gain entry.

From the very start, the council had fastened on a set of discrepancies, or "tensions," in the situation of American children. Myths versus realities, ideals versus behavior, profession versus practice: such were the defining categories of its investigation.

The most obvious and fundamental of these discrepancies appeared to begin from the abiding belief of many Americans that ours is a "child-centered" nation. Typically—we like to think—our children are cherished, protected, nurtured, and offered a field of opportunity unmatched elsewhere in the world. We value children almost to a fault—often, indeed, we "spoil" them—but they are, after all, the hope and the promise of America's future. This rhetoric, these beliefs, are studded through our public professions no less than our private sense of ourselves. And yet, on closer inspection, they yield a painful sense of contradiction. Our public *policies* affecting the young, and the welfare of American children as measured by various social and medical indices, suggest an actual commitment to children substantially lower than that of other industrialized countries. Our patchwork of welfare programs does not amount to support for families with children. Our legal protections for children are meager and inconsistent. Our infant mortality rates are shockingly high. At each of these points of comparison, other countries stand well ahead of us.[6]

In spotlighting this gap between profession and performance, the council hoped to grasp a wedge for change. Recommendations could then be advanced in the name of living up to our national ideals. There was more here, however, than simply a tactical opening. The lack of consistent performance was, and is, a dilemma that must somehow be fathomed. What, in fact, are the *real* attitudes of our country toward children?

A psychiatrist who visited one of the council's early meetings suggested that it consider "whether Americans actually hate children."[7] The proposition seemed stark, and finally implausible, but it could not be ruled entirely out of court. At best, the council came to think, American attitudes toward children are deeply ambivalent. To some extent, of course, ambivalence is built into all intergenerational bonds. Growth and nurturing lead inevitably to separation; the old die, whereas the young live on; the sins of the fathers are visited on their sons, and vice versa. To some extent, the structure of all modern (i.e., industrial) society exacerbates these tensions. In premodern society, children repaid the costs of their care by contributing labor to family enterprise, but now they are massive financial liabilities. Yet with all this accounted for, there is a distinctly American residuum. Somehow, the young in this country seem especially likely to elicit a divided response from their elders.

In the long run, these ambivalent attitudes must be faced, and understood, if American children are to have maximally effective support. But they run so deep, and are so broadly diffuse, as to prevent the drawing of clear connections with substantive policy issues. Other conflicts seem much closer to the visible surface of contemporary American family life, and three of these would eventually become central to the council's "framework" of analysis. In each case, there seems to be great need for policy initiatives in response to real problems and pressures, but the need is obscured, and the problems themselves denied, by the presence of long-standing cultural "myths."

1. *The need for support versus the myth of self-sufficiency.* The prevalent anxiety today about the quality of American family life is not without foundation. Our own families differ from those of our forebears in significant ways, and some of the difference powerfully implicates children. The rapidly increasing participation of women in the labor force is a particularly striking case in point. Among women with school-age children, a clear majority now hold jobs outside the home; the figures leaped from 26% to 54% within the period 1948–1974.[8] Single parenthood is a comparably important phenomenon, given the rising curve of both divorces and out-of-wedlock births.

But the family has been changed in other ways as well. The widely noted loss of its traditional functions is real enough. Work and home no longer have much intrinsic relationship. Education is less and less a domestic affair, and the same can be said of health care. The larger point is that families today must depend on a variety of specialists for vital needs and services. Meanwhile, *expectations* have risen markedly. We want more education, and better medicine, than ever before. And the home itself has been "emptied": parents at work, children at school or elsewhere. The managerial aspects of family life have thus become enormously complicated. The "core report" of the Carnegie Council summarizes the trend as follows:

> In a sense parents have had to take on something like an executive rather than a direct function in regard to their children, choosing communities, schools, doctors, and special pro-

grams that will leave their children in the best possible hands. The lives parents are leading, and the lives for which they are preparing their children, are so demanding and complex that the parents cannot have—and often do not want—traditional kinds of direct supervision of their children.[9]

And yet these needs, these realities run directly against a powerful mythology of family life. The stay-at-home mother, the breadwinner father, and, above all, the family that confidently takes care of itself are enduring images that we all know well. For a family to require help is, by the light of American traditions, to *fail*. When things go awry, the individual family members must somehow be at fault. The price of such beliefs is paid in guilt, in lessened self-esteem, and in disorganized, wholly inadequate programs of social services. In fact, American families have never been the self-sufficient building blocks of society that the myth affirms. And they are becoming less so as time moves along.

2. *The reality of the "stacked deck" versus the myth of equal opportunity.* The council chose the metaphor of a "stacked deck" to designate some of the deepest and most painful dilemmas in American life: the paradox of poverty in the midst of affluence, the crippling burdens of prejudice, the maldistribution of life's chances and rewards. Of all age-groups, children are the most likely to be poor. Moreover, poor children experience, from the beginning of their lives, convergent streams of insult and injury to body and mind. Their health is significantly worse than the average for their age-mates; their rate of mortality is higher; their educational opportunities are far inferior. The odds that they can better themselves are low: According to one study, a child born into the topmost level (highest 10%) of the national wealth hierarchy has a 1-in-3 chance of remaining there, whereas a child born near the bottom (lowest 10%) has a 1-in-250 chance of ever getting there. The blight of poverty is, of course, frequently compounded by the bitter shafts of race prejudice. Black children are four times more likely to be poor than are white children, and native Americans suffer even more terrible odds.[10]

And yet statistics tell only half the story. The emotional scarring, the wounds to soul and spirit, the denial of hope, the warp of expectations—such things cannot be measured quantitatively. Young people willingly learn skills that they know they will be able to use; conversely, they do not learn what seems irrelevant or dysfunctional in terms of their probable life pattern. These judgments, in turn, are based on what they see around them and particularly on the experience of the adults they know best.[11]

The "problems" of poverty and racism are familiar enough by now. Yet their full measure is concealed, and change is hindered, by still another pervasive cultural myth. Our society, we have long believed, is an "open" one; our highest values, our noblest professions, consistently affirm equality, or at least equal opportunity. Where poverty persists, the fault must lie with the individuals involved. Our "solutions" to poverty are fashioned accordingly. Reduced to their common denominator, they amount to this: Change, uplift, "reform" the individual—help him to get started in helping himself.

3. *The perils of the "technological cradle" versus the twin myths of technological progress and the laissez faire economy.* There is no avoiding the impact of technological change on all our lives. Indeed, the scope of this change is so vast as to defy any simple characterization. One can dip into the experience of individuals, or families, at almost any point and find a maze of effects, all stemming in one way or another from technology. The council elected, in its final report, to single out three such points: television, the "new diet," and nuclear energy.

In each instance, thanks to technology, Americans have been offered large apparent gains. Television "expands our horizons," especially those of children. (According to many studies, children now spend more time with this ubiquitous electronic companion than with either their parents or their schools.) The new diet makes food "cleaner" and its preparation more convenient than ever before. Nuclear power holds out the prospect of a dramatic solution to our chronic energy problems. And yet, in each instance, there is vast uncertainty about long-range implications and cause for immediate concern about some of the short-run effects. What is the hidden message to children from television, with its "quick takes" and instant (often violent) solutions for all manner of personal and social problems? What will be the eventual result for their health of the new diet? And what about the risks of nuclear energy—risks that in many respects we cannot even conceptualize?

The country has scarcely begun to take these issues seriously, and once again the lag has much to do with myth. We have, as a nation, a deep and abiding faith in technology as the engine of progress. Our very identity has been shaped by this faith. To be an American has meant participation in a culture of triumphant self-assertion, of mastery within both the natural and the social environment, and the means to this end has been, in large part, a distinctive national gift for technological ingenuity. Many Americans, therefore, now receive the warnings of risk inherent in technology with a special sense of surprise, of indignation, and of panic.

A second myth, also pertinent here, affirms the benign workings of the laissez-faire economy. Free enterprise, and most especially the free choice of consumers, are its central tenets. If television programs are not what they should be, then viewers will switch channels, advertisers will shift their contracts, and producers will come up with something better. If processed foods, laden with cosmetic additives, are dangerous to health, then shoppers will express their preference for something else. If nuclear energy threatens the genetic integrity of future generations, then planners in both the public and the private sectors will react accordingly. These beliefs—they are, in fact, less cognitive than purely instinctive—presume free and informed choice. Yet most of us cannot possibly inform ourselves about such complex matters, and we are far less free than we sometimes imagine. Increasingly our laissez-faire economy is controlled both from within and by government; to depend on "market forces" is therefore patently ill-advised.

The foregoing "tensions" obtained a central place in the discussions of the council, and eventually in its "core report." They directly undergird the opening part of that report—the analysis entitled "Children and Families: Myth and Reality." A second part—"What Is To Be Done?"—offers a variety of policy recommendations. The latter are too numerous and complex for easy summary here, but they embrace the issues of full employment, income supports, flexible work schedules, services to families, health care, and the legal status of children. A concluding chapter discusses ways and means of "Converting Commitment into Politics".

The report was published in September 1977 under the title *All Our Children: The American Family under Pressure*. One additional "council book," *Small Futures: Children, Inequality, and the Limits of Liberal Reform* was published in 1979. One of the "background studies," *Child Care in the Family: A Review of Research and Some Propositions for Policy*, appeared in December 1977; the other two, *Minority Education and Caste: The American System in Cross-Cultural Perspective* and *The Unexpected Minority: Handicapped Children in America*, were published in 1978 and 1980 respectively. This completes the council's publishing agenda.[12]

With the appearance of its final report, the council ceased to exist as a formal entity. However, the members and the staff continue to be active as individuals on behalf of council ideas and recommendations. A last meeting of all those connected with the project was held in Washington, D.C., and interviews were obtained with leading figures in government and private organizations of child advocacy. A small public affairs unit, based in New York, has been helping to funnel speaking invitations to appropriate council members. The director of the project has appeared several times on national television and has testified before a congressional subcommittee on children and youth.

These ongoing activities express a shared conception of the council's work as essentially a form of "consciousness raising." There was never much ground to expect that the publication of books and reports would lead directly to policy change. Rather, the hope is to enhance discussion by concerned citizens everywhere of the forces that actually do, and the principles that ideally should, govern the lives of American children.

III

The appointment of historians to the council and its staff signaled a clear intent on the part of the officers of the Carnegie Corporation. Historical experience—that is, the experience of children in the past—was deemed relevant to the situation of children in the present and the future. The questions of *where* history fitted in and *how* to make it meaningful for the project as a whole were not so easily resolved. Most council members seemed to feel that historical knowledge might offer a useful "background" from which to approach the crucial present-day, policy-oriented issues. And there was a general sense that such knowledge should be deployed at the beginning of the council's work.

So it was that various parts of early council meetings were set aside for consideration of history. How were children viewed in the past? How were they valued? How were they treated? What was the nature of their family experience? How easily, and in what contexts, did they move between their particular domestic settings and the community at large? Questions like these were raised for the entire group by those whose prior experience qualified them for the task. The discussion was distinctly open-ended, and there was

little attempt at connection with the substantive problems of contemporary American children.

There were other manifestations of the same impulse. The
director of the council began, at an early stage, an extensive study
of attitudes toward children in the past. (The result was a manuscript of several hundred pages.) A staff member with particular
training and interest in history was asked to review the outcome of
the various White House conferences on children from 1900 to the
present. (The same person would subsequently undertake to research the history of day care.) Other staff members began to
investigate poverty from a historical perspective and the changing
workroles of women during the past two centuries. The early
emphasis on nutrition sparked a modest foray into food history. In
fact, all of these studies fell by the wayside as the council's work
reached a more advanced stage. A few may well find their way into
print at some future date, but their immediate priority in terms of
council goals did not prove in the end to be high. Were they,
therefore, entirely without significance? On the contrary, they appear to have entered, in numerous small ways, the larger flow of
ideas that the council was gradually generating. But this result is
clearer in retrospect than it was at the time.

One particular meeting, midway through the project, elicited a
substantial exchange of views on the "uses of history." The director had circulated several "pre-draft" chapters from his ongoing
study of children in the past, and the council was asked to react.
One member—the only real skeptic in this connection—doubted
"the relevance of history to the problems of children and families
in American society now. . . . People do not decide to do things by
taking account of historical findings; instead they look at economics, politics, and the immediate social context . . . in order to
determine how and what to do." (This, and succeeding quotations,
are taken from the meeting minutes.[13]) But others, from both the
council and the staff, took an opposite view. History was "important . . . as a means of dispelling fears of change." History would
be helpful "in identifying the needs of children, and the sources of
those needs." History might serve to "unveil certain nostalgic
myths" that had traditionally frustrated or distorted policy efforts.
History would teach by example, spotlighting "those programs
[which had been] tried time and time again with almost no success." In effect, the council reaffirmed a generalized interest in
history, even while setting aside specific inquiries about the past.

There was one additional sign of affirmation, indirect and

unintended, but hardly unimportant. Repeatedly, the council historian noticed his policy-minded colleagues opening some new line of discussion with a brief reference to the past. An argument for expanded day-care facilities would be introduced with the thought that day care was not, after all, so very new; the "day nurseries" of the nineteenth century and the federally sponsored preschool centers of the World War II era could be seen as approximate precedents. A comment about the need for additional "family services" might start from a notion that the mobility of Americans, in the geographical sense, is greater now than ever before. A discussion of the impact of divorce on small children would explicitly presume that family separations were rare long ago. This tendency amounted, indeed, to an almost instinctive impulse—an "impulse to the historical preface" (so to speak). For many council members, it was as if the record of the past conferred a certain legitimacy on ideas about the present.

These prefaces were not always accurate—at least by the light of academic research—and some of them had no necessary bearing on the policy positions with which they were linked. If American children of the 1970s can benefit from day care, does it matter much whether similar arrangements have existed in the past? If contemporary families truly need expanded "services," must policymakers decide whether or not an increased rate of geographic mobility is the cause of this need? If millions of children are adversely affected by the rising divorce rate, what difference does it make (for policy) that such occurrences may have been less common in the past? But, within the context of council discussion, it seemed immaterial whether such connections could actually be demonstrated. Somehow, the preface was its own justification.

Meanwhile, the larger flow of the council's work was reaching toward history in ways far more substantial, if less immediate and obvious. The priorities, the assumptions, the "strategic decisions" described earlier in the chapter carried important historical implications. The council would adopt an "ecological" approach, as opposed to that of the "problems list." The former was by definition multidimensional, and one of the dimensions, quite clearly, was time. The council would interpret its charge to comprehend all children, not simply the most disadvantaged, and would key its investigation to continuing trends and tendencies, not a utopian vision of an ideal world. Such notions had little meaning apart from comparisons of the past with the present.

Most of all, the family focus reached out to history—indeed depended on history. It is not impossible, in the last decades of the 20th century, to imagine that the family system itself is disintegrating. Divorce, illegitimacy, the burgeoning technology of contraception, a new brace of "countercultural" standards and values— these undoubted landmarks of contemporary life suggest to some that families are virtually outmoded. If that were truly so, policy discussion about children would have to begin from new and "radical" premises. The council, however, elected to think otherwise, and history was strongly implicated in its decision. The record of the past, at least by this reading, powerfully displays the durability of the family system. Concern for the family has reached high levels before; yet its inner structure, and its fundamental integrity, have remained largely intact for centuries. Change it has indeed experienced, and further changes can be expected. There are few grounds, however, for predicting its imminent demise or the *kinds* of change that would lead to something genuinely new.

The council returned often to the problem of cultural ambivalence toward children, and here, too, history was asked to supply understanding. The roots of this ambivalence invited extended exploration—more extended than the council could reasonably undertake and more, certainly, than can be attempted here. But a brief sketch may serve to exemplify the mode and manner of such discussion.

Consider as a way of approaching the problem the special situation of one type of American family—the immigrant family. It is a commonplace of social science research that immigrant families are liable to inner stress and tension.[14] For one thing, the children—those who are raised here—may prove to have certain advantages over their Old World parents. They are, in general, more adaptable: They learn the language more quickly, as well as the mores, the values, the prevalent *nuances* of social style. At some points, the parents may actually have to depend on their children for advice and assistance of a sort that has traditionally flowed the other way. Is it not plausible that these reversals may seem, at least in part, unwelcome to the parents?

Here, then, is one potential source of resentment in the older generation, and there are other things that may trouble them as well. *Guilt* is a factor that seems intrinsic to immigrant experience, and directly pertinent to family life. The parents feel that the decision to migrate has placed all members of the family at risk; in particular, it has robbed the children of their proper birthright in a stable and rooted community. Of course, many immigrants have justified their coming in terms of the gains expected for the children, but at some level they recognized that there was also a loss. This tacit recognition has prompted, in turn, a further response, a compensatory pledge that they as parents would subordinate their own interests to the future prospects of the young. Here, incidentally, one encounters the *positive* side of the cultural ambivalence toward children. Enormous sacrifices have, in reality, been made by many parents for many children in this country; to this extent, the myth of our "child-centered" society finds confirmation. And yet there is still another twist. The immigrant parent who has sacrificed all understandably expects a return. "My own life has been foreshortened," he seems to say to his children, "but you will live for me; the luster of your success will redeem my bitter struggles." This, most assuredly, is an open door to intrafamilial tension of various kinds, and to potentially deep disappointment.

The foregoing sequence has been presented as specific to immigrant families, but, in truth, many of the same themes can be associated with American family life more generally. Immigration has been a central and dynamic element in our history from the very start; the Puritan settlers were, after all, the first immigrants. Moreover, millions of Americans who are certifiably "native stock" have shared with their foreign-born neighbors the experience of sudden geographical movement. In short, they are migrants, if not immigrants. The motion in *their* lives has raised some of the same expectations, and fueled some of the same tensions, that have often characterized immigrant families.

To these pressures from social experience was added the impact of ideological forces, dating back well over a century. Beginning in the period 1800–1850, for example, Americans developed a special and highly optimistic view of social evolution—their own version of the "idea of progress." The country had made a successful revolution and come through a difficult period of adjustment in the generation or so after independence. The westward movement was in full swing; the economy was expanding dramatically; the political system appeared to have a "vanguard" status in relation to other parts of the world. Americans of all sorts were congratulating themselves on their "go-ahead spirit" (a familiar period cliché), and they truly did believe that they had launched a wonderful experiment in human improvement.

According to this view, the present was inherently better than the past, and the future was certain to be better than the present. And who were the representatives of the future? Who else, of course, but the nation's children. In a sense, the younger generation could claim a moral edge on the older one, as the flow of history moved along. And to the extent that the young asserted this claim, family relations, or even social relations generally, were unsettled. In premodern society, the superiority of age had been everywhere recognized, but now much of that was turned around.

Another ideological innovation of the same historical period lay in the meaning and emphasis given to human achievement. This was part, to be sure, of a larger climate of "individualism," in which the worth and talent of each person became a touchstone of social value. But the key development was the so-called cult of success. According to the premises of this cult, life was a race, open to talent and with tangible prizes for the winners. The central figure here was the "self-made man," the man whose success was attributable solely to his personal strivings.

The implications of these values for the relation of parents and children were profound. It was accepted from this time forward that the way for a young person to move ahead was to leave the world of his parents and strike out on his own. (For "young person," we should probably substitute "young man," since these formulas were applied chiefly to males.) Ties to family, attachment to the community of his birth—all this could only hold him back. Would his parents' experience at least serve as a guide, a model, as he made his way in the wider world? At best, the answer was ambiguous. The ethical teachings of parents might prove a valuable "rudder" for him as he veered hither and yon—and nautical metaphors were common in this connection—but other aspects of the parental experience would be irrelevant. He would choose his own occupation, his own friends and colleagues, his own place of residence, wherever the main chance seemed to beckon. In short, getting *ahead* meant getting *away* from one's parents.

And there was more. How would the young person's achievement be measured? His achievement was measured, his "success" evaluated, by comparing his ultimate social position with the point from which he had started out. In short, success meant surpassing one's father. So it was that a deep strain of competitiveness entered the relation of parents and children in nineteenth-century America. It has remained with us, indeed in us, ever since.

This is a fair specimen of council inquiry specifically embracing historical circumstance. It begins from a question or problem founded in present-day experience. (Whence comes the characteristic American ambivalence toward children?) It proceeds at a high level of generality and requires no direct acquaintance with "primary" data. It draws, at least implicitly on the research of whole legions of scholars. And it suggests no solutions, carries no direct implications for policy, since it purports to express the accumulated experience of an entire culture.

Much the same could be said of the place of history in the council's final report. Historical ideas figure quite prominently in the opening sections of that report, in the "analysis" that precedes the offering of particular recommendations and findings. Recall the quartet of cultural "myths" presented in counterpoint to problematic social "realities." The myth of self-sufficiency, for one, is clearly a legacy from the past. Much depends on the growth, especially in the nineteenth century, of an inside–outside view of the family in relation to the larger community. The modernization of American society raised deep, and often unadmitted, anxieties among many of those most centrally involved. The new world of commerce and industry seemed chaotic, impersonal, and amoral; the home, by contrast, was idealized as a bastion of traditional virtue. The boundaries were maintained with ever-increasing vigilance, and "self-sufficiency" became the means to this vital end. (The echoes of this view, with its shrill emphasis on boundary maintenance, can be heard in the opposition to the child-care legislation of 1971 discussed above. The hallowed division of sex roles (breadwinner and housewife) was similarly founded. The "cult of true womanhood," as one historian has called it, was founded on a belief that maternal care alone guaranteed the safety and virtue of children.

The myth of equal opportunity, which prevents our full confrontation with structural poverty and racism, also dates from the early nineteenth century. And, at the outset, it contained at least a kernel of truth. From well before the War of Independence, American society was more homogeneous in its social and economic structure than was its mother country overseas. And the Revolutionary ideology helped to convert these "facts" into values. After 1800, as the national economy entered a "take-off" stage, there were indeed unusual opportunities for enterprising (and lucky) folk to get ahead. Perhaps the single strongest boost for the myth of egalitarianism came from the creation of a free, and later compulsory, public school system: Schools would be the starting line from which all competitors might equally begin. That this mythology has survived so long, and in the face of so much evidence to the contrary, seems nothing short of incredible. But it carries the dreams, and softens the disappointments, of whole generations of Americans.

The historical derivation of the final pair of myths—those that shroud our "technological cradle"—is so transparent as hardly to need comment. The myth of technological progress encapsulates more than a century of extraordinary economic growth, of manifold "invention," and of ever-expanding comforts and "conveniences." For most Americans, technology was, and remains, the tangible expression of the confidence, the resourcefulness, the practical bent that together have created our modern "life-style." The myth of laissez-faire has a not dissimilar set of roots. The energy of individual producers needed only to break free of traditional restraints in order to yield its most bountiful possible result. The public interest was the sum of many personal interests, no more and no less. English writers of the seventeenth and eighteenth centuries, like John Locke and Adam Smith, had forged the philosophical underpinnings of laissez-faire, but American *do*ers of the nineteenth century effected its social application.

The later sections of the council's final report offer fewer, and markedly smaller, openings to history. Specific recommendations for policy look naturally to the present and the future. Nonetheless, these chapters are dotted at intervals with brief references to the past, which do bear some consideration.

Most such references can be divided among three different categories. One group invokes history on behalf of policy. (Hence, e.g., the following: "It is consistent with almost any interpretation of American traditions and values to insist that parents' efforts to secure an adequate income for their family not be thwarted by the brute fact that there are no jobs for them."[15]) A second group serves to highlight particular features of the present. In some cases this is achieved by way of contrast. (E.g., "'Services' is the catchall term for many of the kinds of help that parents use now that life is not the simple family affair it was on eighteenth-century farms."[16]) In others, the emphasis is on continuity—that is, the power of a linear tendency over time. ("Barring a dramatic reversal in the trend that decreased the proportion of self-employed workers in the labor force from 20 percent in 1940 to about 8 percent in 1970, changes in the working conditions of the 92 percent who work for others will have the greatest impact."[17]) A third group of these references presents the past in a darker light. Here the emphasis falls on the burden of inherited practices and principles, whose very longevity belies their worth. (Thus, "For far too long we have applied to the job problem the same old solution—uplift and reform the individual—that has proven so difficult to achieve in other realms."[18]) In short, the "problem" and its purported "solution" are both of considerable duration; hence the latter must be deemed ineffective.

In all the above-mentioned instances, the purposes served are essentially tactical. The substance, for the most part, is left rather vague; the tone is usually rhetorical. There is little intrinsic relation between the cited historical circumstance and particular policy recommendations. Indeed, we are back to the matter of "prefaces," and it is no coincidence that many of these references appear at the beginnings of paragraphs or of the larger textual subdivisions.

IV

What, now, can be said of the role of the historian on the council? And how can the council's experience be used to illuminate the uses of history, and of historians, for similar policy investigations in the future? In turning to these questions, it seems appropriate to speak in the first person. There is no way finally to escape the subjective aspect of my participation in the work of the Carnegie Council. And I cannot at this point be sure how much my role, and my conclusions about that role, may parallel the experience of other historian-participants in projects of a similar nature.

I must admit, in the first place, to some unease about my contribution to the council virtually throughout. It seemed from the start that most of my colleagues were more immediately and tangibly connected with the task than I was. In effect, they represented particular constituencies—disturbed children, poor children—by virtue of much professional experience; or else they represented some distinct expertise—pediatrics, law, economics—with direct bearing on the lives and prospects of children. My constituency, if such it can be called, was the children of the past, and they were clearly beyond reach. My expertise seemed inherently less practical, more diffuse, and more esoteric than the others. None of my fellow council members wished to make me feel uncomfortable about this—quite the contrary—but nevertheless I experienced my role as requiring special definition.

Several possibilities emerged, and to varying degrees I tested them all. The project began, as noted earlier, with efforts to use history as "background." My own part in these efforts was necessarily prominent: the preparation of a general paper on the history of family life, and extended comment on the history of childhood specifically. The immediate results were disappointing, to say the least. My rendering of history seemed to lead nowhere in particular, and the reaction around the table was understandably diffident. Such background presentation seemed vaguely comparable to the tunes piped into supermarkets and airline terminals. To this point, it could be said, mine was the role of *the historian as music maker.*

The background had only a limited claim on the council's attention, and quite soon I was searching for new forms of participation. The aforementioned "impulse to the historical preface" apparently offered a way. I could, and did, become an occasional spokesman on behalf of historical accuracy. From time to time, as the comments of my colleagues created opportunity, I would correct the record—*the historian as gadfly.* "But wait," I would interrupt, "it really wasn't quite that way; recent studies have shown . . ." At the same time, I would suppress my sense that such historical corrections were often irrelevant to the policy questions immediately at hand.

But this, too, was a function of minimal reward, both from my individual standpoint and for the project as a whole. And increasingly I drifted toward the position of a "generalist"—*the historian as intelligent layman.* I sought to understand at least something of the specialities that my colleagues variously represented and to assist in melding them into a larger viewpoint. Historians are practiced in the arrangement of facts and the construction of arguments; so I might serve as a sounding board against which others might bounce their ideas. Historians are also practiced in writing, and when the final report had been drafted (mostly by other hands) I spent several days pouring over it with a red pencil, as a self-appointed editor.

By this time, however, I had actually done some service from out of my own specialty—*the historian as historian.* And, curiously, this happened almost without my noticing it. As the council fashioned its analysis of children's experience in the present, questions about the past regularly popped into view. It seemed important, above all, to locate the origins of the various "myths" to which the final report would assign such significance. Here I could contribute directly, albeit in a rather piecemeal and unsystematic way.

The belated and roundabout entry of historical ideas into the mainstream of council work implies a general point of some interest. Perhaps we erred in approaching history in the way it is usually approached by scholars—that is, moving from the remote past, to the more recent past, to the present. Indeed, I suspect that history's relevance to policy is most easily appreciated along a route that *reverses* chronology—that is, from present to past. And if this is true, then historical investigation belongs to a later, rather than an earlier, stage in the work of a group such as ours. One needs questions to put to history, and questions arise only as policy itself receives shape and substantive definition. To proceed otherwise, as the council initially did, is to invite discussions that become "academic" in the literal sense.

The largest of the questions that the council put to history reflected its overall "ecological" orientation. Effective advocacy for children, so we concluded, requires far-reaching social change. Hence we asked: what elements of our culture facilitate change, block change, distort and inhibit change? Which of our traditions can be marshaled on the side of new policy initiatives, and which serve merely to maintain the status quo? The council sought to make these inquiries by counterpointing inherited "myth" with immediate, and problematic, "reality." As such, our analysis reached out to one branch of history in particular: the history of values, of custom, of ideology.

It is worth noticing, in conclusion, what the council did *not* attempt by way of historical investigation. The evolution of cultural "myth" is an important matter for any project concerned with policy change, but it hardly covers all relevant possibilities. The council might, for example, have used history to reflect more fully on its own place and function. What have similar projects achieved in the past? What does the historical record suggest about professional advocacy for specifically targeted groups, such as children, within the larger population? Are there often unintended results that may rival in importance the ostensible "findings" and recommendations? Can one discover—again, from historical inquiry—meaningful implications in the fact of foundation sponsorship, or in the social position of the project members, or in the inevitable dealings with government, the media, and other institutional structures?[19] Answers to any or all of these questions might possibly have altered the council's view of itself and its subsequent mode of operations.

There is, moreover, a second route toward the past that the council also avoided. It may well seem odd to some readers of *All Our Children* that the book presents so little history *of* policy for and about children. Health programs, nutrition programs, the whole vast record of public education—are there no lessons to be learned here that might guide policy for the future? What of the history of the Children's Bureau, of local agencies for child and family services, and of private philanthropies in this same connection? And if indeed the record of policy specifically for children seems at many points rather thin, what might be learned from studying past programs for other groups (e.g., the elderly)?

Perhaps with a different membership, a different staff, and, in particular, a different historian, the council would have tried such approaches. But not *this* council, as originally constituted. An extended effort to reflect on the context of our own work—a process of institutional introspection—would, I suspect, have made most of us quite impatient. An agenda of research around the details of past policy would have seemed inordinately fussy and esoteric. In fact, few studies of the latter sort have been made by anyone, so the council would have had to start almost from scratch. Furthermore, historians themselves are skeptical about using the past to read lessons for the future. The context shifts, the characters come and go, the equation of cause and effect is never the same—such, at least, is our usual presumption.

Still, it is far too early for closure on any of these questions. Historical inquiry and policy formation make a new, awkward, and necessarily uncertain tandem. But practice may yet bring greater synchrony and increasingly substantial results.

NOTES

1. Alan Pifer, "Preface," in Kenneth Keniston and the Carnegie Council on Children, *All Our Children: The American Family under Pressure* (New York, 1977), p. ix.

2. The members of the Carnegie Council were the following: Kenneth Keniston, Chairman, psychologist; Catherine Foster Alter, social worker; Nancy Buckler, teacher of disturbed children; John Demos, historian; Marian Wright Edelman, lawyer; Robert J. Haggerty, pediatrician; William Kessen, psycholo-

gist; Laura Nader, anthropologist; Faustina Solis, social worker; Patricia McGowan Wald, lawyer; and Harold W. Watts, economist.

3. On the congressional legislation, see H.R. 19362, as noted in U.S., Congress, House of Representatives, *Congressional Record*, 91st Cong., 2d sess., 1971, 116, pt. 24:32820–3281. For testimony about the bill, consult *Hearings before the Select Subcommittee on Education of the Committee on Education and Labor* (Washington, D.C., 1970), *passim*. For congressional debate on the subject, see *Congressional Record*, 91st Cong., 2d sess., 1971, 117, *passim*.

4. Elinor Guggenheimer, "Look Here, Mr. Nixon," *New York Times*, 21 December 1971, p. 37.

5. The main elements of this discussion are summarized in Carnegie Council on Children, Minutes of the Third Meeting, 1–2 December 1972. Carnegie Corporation, New York, N.Y.

6. On the deficiencies in current policy for children, see Keniston *et al.*, *All Our Children*, pp. 78–80.

7. See Carnegie Council, Minutes of the Second Meeting, 3–4 November 1972, pp. 4–6.

8. Keniston *et al.*, *All Our Children*, p. 4.

9. *Ibid.*, pp. 12–13.

10. *Ibid.*, pp. 32–33.

11. This point is explored at length in one of the council's background books, John U. Ogbu, *Minority Education and Caste: The American System in Cross-Cultural Perspective* (New York, 1978), especially chaps. 1, 5, 6.

12. The full citations for those works not already cited are as follows: Richard Delone, *Small Futures: Children, Inequality, and the Failure of Liberal Reform* (New York, 1978); Alison Clarke-Stewart, *Child Care in the Family: A Review of Research and Some Propositions for Policy* (New York, 1977); John Gliedman and William Roth, *The Unexpected Minority: Handicapped Children in America* (New York, 1979).

13. Carnegie Council, Minutes of the Sixteenth Meeting, 13–14 December 1974, pp. 3–4.

14. The fullest and most insightful exposition of the argument summarized in these paragraphs is found in Oscar Handlin, *The Uprooted*, 2d ed. (Boston, 1973).

15. Keniston *et al.*, *All Our Children*, p. 85.

16. *Ibid.*, p. 135.

17. *Ibid.*, p. 124.

18. *Ibid.*, p. 91.

19. A provocative book by Christopher Lasch sketches one set of parameters for such discussion. See his *Haven in a Heartless World* (New York, 1977).

Index